Geothermal Power

Carla Mooney

Energy and the Environment

ReferencePoint
Press®

San Diego, CA

© 2012 ReferencePoint Press, Inc.
Printed in the United States

For more information, contact:
ReferencePoint Press, Inc.
PO Box 27779
San Diego, CA 92198
www.ReferencePointPress.com

Picture credits:
Cover: iStockphoto.com
Maury Aaseng: 31–34, 46–48, 60–62, 75–77
© Corbis/Guido Cozzi: 9
iStockphoto.com: 17

LIBRARY OF CONGRESS CATALOGING-IN-PUBLICATION DATA

Mooney, Carla.
  Geothermal power / Carla Mooney.
    p. cm. — (Compact research series)
  Includes bibliographical references and index.
  ISBN-13: 978-1-60152-162-0 (hardback : alk. paper)
  ISBN-10: 1-60152-162-6 (hardback : alk. paper) 1. Geothermal resources. I. Title.
  GB1199.5.M66   2011
  333.8'8—dc22

                                                                2010053307

Contents

Foreword

> **❝Where is the knowledge we have lost in information?❞**

—T.S. Eliot, "The Rock."

A s modern civilization continues to evolve, its ability to create, store, distribute, and access information expands exponentially. The explosion of information from all media continues to increase at a phenomenal rate. By 2020 some experts predict the worldwide information base will double every 73 days. While access to diverse sources of information and perspectives is paramount to any democratic society, information alone cannot help people gain knowledge and understanding. Information must be organized and presented clearly and succinctly in order to be understood. The challenge in the digital age becomes not the creation of information, but how best to sort, organize, enhance, and present information.

ReferencePoint Press developed the *Compact Research* series with this challenge of the information age in mind. More than any other subject area today, researching current issues can yield vast, diverse, and unqualified information that can be intimidating and overwhelming for even the most advanced and motivated researcher. The *Compact Research* series offers a compact, relevant, intelligent, and conveniently organized collection of information covering a variety of current topics ranging from illegal immigration and deforestation to diseases such as anorexia and meningitis.

The series focuses on three types of information: objective single-author narratives, opinion-based primary source quotations, and facts

4

and statistics. The clearly written objective narratives provide context and reliable background information. Primary source quotes are carefully selected and cited, exposing the reader to differing points of view. And facts and statistics sections aid the reader in evaluating perspectives. Presenting these key types of information creates a richer, more balanced learning experience.

For better understanding and convenience, the series enhances information by organizing it into narrower topics and adding design features that make it easy for a reader to identify desired content. For example, in *Compact Research: Illegal Immigration*, a chapter covering the economic impact of illegal immigration has an objective narrative explaining the various ways the economy is impacted, a balanced section of numerous primary source quotes on the topic, followed by facts and full-color illustrations to encourage evaluation of contrasting perspectives.

The ancient Roman philosopher Lucius Annaeus Seneca wrote, "It is quality rather than quantity that matters." More than just a collection of content, the *Compact Research* series is simply committed to creating, finding, organizing, and presenting the most relevant and appropriate amount of information on a current topic in a user-friendly style that invites, intrigues, and fosters understanding.

Geothermal Power at a Glance

Heat from the Earth

Geothermal energy is generated in the earth's core. Temperatures hotter than the sun's surface are continuously produced inside the earth by the slow decay of radioactive particles, a process that happens in all rocks.

Uses of Geothermal Energy

Geothermal energy can be used to generate electricity, operate heat pumps, supply space heating, and provide hot water.

Where Geothermal Energy Is Found

The most active geothermal sites are found where earthquakes and volcanoes are concentrated, specifically in an area called the Ring of Fire that encircles the Pacific Ocean.

Geothermal Power Generation

The United States leads the world in electricity generation with geothermal power, but geothermal power plants produce 0.4 percent of total electricity generated.

Environmental Benefits

Geothermal energy is one of the cleanest energy sources, with lower emissions and pollution than fossil fuels. Geothermal also uses less land than solar panels and wind farms and does not create harmful toxic waste like nuclear power plants.

Limited Availability

Current technology limits geothermal resources to regions near underground geothermal reservoirs. Outside of those areas, geothermal energy is often not available or economical.

High Costs

Drilling wells and the construction of geothermal plants require a high up-front investment, which is one barrier to expanded use of geothermal power.

Risk of Earthquakes

The process of extracting geothermal energy from the earth can trigger earthquakes. To date, the seismic activity has been small, but some fear that deeper drilling will trigger a damaging quake.

Promising Future

With new technologies in development to reduce the cost and expand access to geothermal energy, the future of geothermal appears bright.

Overview

66Geothermal energy could play an important role in our national energy picture as a non-carbon-based energy source. It's a very large resource and has the potential to be a significant contributor to the energy needs of this country.99

—Nafi Toksöz, professor of geophysics at the Massachusetts Institute of Technology.

66There remains a large gap—an energy gap between the reliable energy that we need—and the renewable energy that we want.99

—John Barrasso, US senator from Wyoming.

Geothermal Energy Heats Iceland

Surrounded by the frigid waters of the North Atlantic Ocean, Iceland digs deep below the earth's surface for heat and power. Heat from the earth's core, called geothermal energy, powers Iceland's capital city, Reykjavík. Pipes draw hot water from underground springs to provide nearly 90 percent of the city's heat and hot water. The spring water supplies hot tap water and heats buildings, swimming pools, and greenhouses. In the winter it heats sidewalks and roads that accumulate snow. In addition, steam from hot underground springs generates approximately 25 percent of Iceland's electricity needs. Using geothermal energy has reduced Reykjavík's dependence on fossil fuels, transforming it from a polluted city to one of the cleanest in the world.

Icelanders revel in the warm water of an outdoor geothermal spa. Water heated deep underground not only enhances recreational activities, it also provides heat and power for homes and businesses through geothermal power plants such as the one pictured in the background.

Iceland's geographical location enables it to harness geothermal energy. The country sits along the Mid-Atlantic Ridge, a deep-sea mountain range with a high concentration of volcanoes. The volcanic area holds shallow magma reserves that heat enormous underground reservoirs of water. Icelanders have learned to transform thermal heat into energy, virtually eliminating the country's need for fossil fuels for heat and electricity.

Despite widespread use of geothermal energy, Iceland may only be scratching the surface of its geothermal resource. "It's been estimated that by conventional use of geothermal, the available power in Iceland could be on the order of 20 to 30 terawatt-hours per year. Currently we're producing maybe four terawatt-hours per year,"[1] said Ólafur Flóvez, general director of Iceland Geosurvey, a government research institution.

What Is Geothermal Energy?

More than 4,000 miles (6,437km) below the surface, the earth's core continually produces heat, mainly from the slow decay of radioactive materials such as uranium and potassium. The intense heat, which can reach temperatures over 9,000°F (4,982°C), flows continuously from the core outward to the surrounding layer of rock called the mantle. As the heat transfers to the mantle, the rock melts and becomes magma. In some cases the hot magma reaches the earth's surface as lava and forms volcanoes. Most of the time, however, magma stays below the earth's crust, where it heats surrounding rocks.

Groundwater seeps into the earth through cracks and porous rock and is heated by the hot rocks. In some places the water collects in underground reservoirs that can reach temperatures of 700°F (371°C). Sometimes, the hot water and steam escape through faults and cracks in the earth's crust, pooling in hot springs or bursting into the air as a geyser.

Using geothermal energy has reduced Reykjavík's dependence on fossil fuels, transforming it from a polluted city to one of the cleanest in the world.

To use geothermal's hot rocks, steam, and water for energy, the thermal energy must be converted into electricity or used in other ways. Developers drill wells into geothermal reservoirs. The force from high-pressure water and steam can be used to spin turbines and generate electricity. Hot steam and water can also be used directly to heat buildings, pools, greenhouses, and sidewalks.

The potential from geothermal energy is enormous. According to a US Geological Survey report, "Even if only 1 percent of the thermal energy contained within the uppermost 10 kilometers of our plant could be tapped, this amount would be 500 times that contained in all oil and gas resources of the world."[2]

Where Geothermal Reservoirs Are Found

Most geothermal reservoirs are located deep underground, with no visible clues showing above ground. Underneath the surface, the earth's crust is made of several enormous plates, which constantly move very slowly. The areas where these large plates meet, collide, and sometimes slide beneath each other are plate boundaries. Hot magma rises through the crust near these plate boundaries, sometimes forming volcanoes. One of the most active geothermal areas in the world is called the Ring of Fire because of its many volcanoes. Along the Ring of Fire, earth's oceanic plates collide and slide beneath the continental plates, creating volcano ranges and bringing magma close to the surface. The Ring of Fire borders the Pacific Ocean and includes the South American Andes, Central America, Mexico, the western United States, Alaska, and western Canada, and parts of Russia, Japan, the Philippines, Indonesia, and New Zealand.

Geothermal reservoirs and volcanoes also form in areas like Iceland and Kenya's Rift Valley, where the plates move apart and hot magma rises to fill the gap. Other active geothermal sites occur at hot spots, where the earth's crust is thin enough to allow hot magma to rise to the surface. The Hawaiian Islands are a string of volcanic islands formed over a geothermal hot spot.

Although the highest-temperature geothermal reservoirs are usually found near plate boundaries or hot spots, other forms of geothermal energy can be found all over the world. From 10 feet (3m) to a few hundred feet below the earth's surface, there is a steady supply of milder heat that can be used for direct heating. In addition, a vast amount of hot dry rock exists about 2 to 6 miles (4 to 10 km) below the earth's surface. If scientists can develop technology to capture this heat, geothermal energy may be available worldwide.

Uses of Geothermal Energy

Geothermal energy can be used in several ways. High-temperature water or steam from a geothermal reservoir spins turbines to generate electricity. Once the geothermal energy has been converted into electricity, power lines transmit it across large distances to heat homes and businesses and to power appliances.

If the geothermal reservoir is not hot enough to generate electricity, it can be used in direct heating applications. Direct-use systems typically use local geothermal water because it is difficult to transport heat over large distances. Pumped from the ground, the warm water circulates through a radiator or another type of heat exchanger that extracts as much heat as possible. The cooled water is then disposed of or injected back into the ground, where it flows through the hot rocks and reheats. This type of direct-piped hot water system heats almost all of the buildings in Reykjavík, Iceland. In addition to heating, direct-piped geothermal water can be used for bathing, washing, and other similar uses.

Several industrial applications also use geothermal energy. In France direct-piped hot water is used in agriculture and fish-farming projects. Shallow reserves of lower-temperature spring water (70°F to 300°F, or 21°C to 149°C) heat greenhouses, de-ice roads, heat fish farms and spas, and heat air that is used to dry out fish, fruits, and vegetables.

> **One of the most active geothermal areas in the world is called the Ring of Fire because of its many volcanoes.**

In contrast to direct-piped systems and power plants, geothermal heat pumps cool and heat buildings without being near geothermal reservoirs. A few feet below the earth's surface, the ground stays a constant temperature of approximately 40°F to 50°F (4.4°C to 10°C) year-round. The geothermal heat pump sends air or a liquid such as antifreeze through underground pipes. Because heat always flows from warmer to cooler areas, the heat from the ground transfers to the air or liquid in the pipes and circulates through the building. During the summer months the pump works in reverse by transferring the heat from the air inside the house to the cooler ground. A heat pump requires energy to run, but much less than traditional furnaces and air conditioners. In areas with extreme temperatures, geothermal heat pumps are one of the most energy-efficient heating and cooling systems.

Development of Geothermal Energy

Geothermal energy has been used for centuries. Ancient Romans used geothermal water to heat buildings in Pompeii. Native Americans bathed in hot mineral springs and believed in their healing power. The Maori people of New Zealand used geothermal water for cooking.

In 1892 the use of geothermal energy became more sophisticated when the town of Boise, Idaho, developed the world's first modern district heating system. Residents piped water from nearby hot springs to heat the town's buildings. Within a few years the Boise's district heating system heated about 200 homes and 40 businesses. Today Boise has 4 district heating systems. The systems heat more than 5 million square feet (464,515 sq m) of residential, commercial, and government space. Following Boise's lead, several communities around the world have built systems that use the earth's hot water to heat homes and businesses.

> " **High-temperature water or steam from a geothermal reservoir spins turbines to generate electricity.** "

Until the twentieth century geothermal energy could only be used locally because too much heat was lost during transport over long distances. In 1904 workers in Larderello, Italy, experimented using geothermal steam to produce electricity. They drilled a hole into the ground and inserted a pipe into a large geothermal reservoir that produced hot, dry steam. The pressurized steam shot up the pipe and ran the electricity-producing equipment.

Geothermal Power Around the World

Today the United States leads the world in generating electricity from geothermal energy. According to the Energy Information Administration, geothermal power plants in the United States produced 15.2 billion kilowatt-hours of electricity in 2009. Even so, electricity generated by geothermal energy is only about 0.4 percent of the total amount of electricity generated annually in the United States. In addition to the power plants, more than 300,000 homes, schools, and offices use geothermal heat pumps to keep temperatures constant and comfortable.

Worldwide, several countries are pursuing geothermal energy. As of 2010, 24 countries had geothermal power plants. Next to the United States, the Philippines was the world's second-largest producer of geothermal power in 2010, generating 1,904 megawatts, or 18 percent of the country's electricity. In El Salvador geothermal power plants produce 26 percent of the country's electricity. In addition, district heating and direct-use geothermal applications are found in even more countries, including Russia, China, and Sweden.

How Geothermal Energy Is Obtained

Geothermal heat pumps use the earth's temperature near the surface and do not need the water or steam of a geothermal reservoir. As a result, they only require shallow drilling in the earth's crust. Accessing a geothermal reservoir to run a power plant, however, is considerably more complicated and may require drilling miles below the earth's surface.

Geothermal power development starts with exploration to locate a usable geothermal reservoir. Scientists and engineers use geological, electrical, magnetic, geochemical, and seismic survey tools to help them locate geothermal reservoirs. When developers believe they have located a reservoir, they drill an exploratory well to confirm its location. Exploratory wells also measure the reservoir's subsurface temperature and flow rates to determine if it will be a good producer of geothermal energy.

Once a reservoir has been proved, the site is developed for either electricity generation or direct use. Engineers drill production wells into the earth, sometimes over 2 miles (3.2km) deep. Hot water and steam shoot up the wells naturally or are pumped to the surface. High-temperature water and steam (250°F to 700°F, or 121°C to 371°C) run turbines and generate electricity at a geothermal power plant, while lower-temperature water and steam (70°F to 300°F, or 21°C to 149°C) are pumped for direct-use applications.

Geothermal Power Plants

Geothermal power plants convert the thermal energy in underground reservoirs of hot water and steam into electricity. Unlike fossil fuel power plants, geothermal power plants do not burn fuel or have smoky emissions. Instead, geothermal power plants use hot water and steam to run electric generators. Geothermal power plants can then send electricity via power lines to users at greater distances.

The design of a geothermal power plant depends on whether the geothermal reservoir produces dry steam, wet steam, or hot water. The simplest type of geothermal power plant is a dry steam plant, like the first geothermal power plant in Larderello. The steam flows through pipes to an electricity-generating turbine, where the force of the steam spins the turbine. Afterward, the steam passes through a condenser, where it is cooled and condensed into water. The plant then returns the water into the ground, where it will be reheated and recycled by the earth's hot rocks. Although the most efficient type of plant, dry steam plants are rare because few geothermal reservoirs produce hot, dry steam. The Geysers dry steam reservoir in Northern California is the largest known dry steam field in the world.

The most common geothermal power plant is a flash steam system. This type of plant depends on geothermal reservoirs that hold hot water instead of steam. As the hot water rises through the pipes and is released from the pressure of the deep reservoir, some of it boils and turns to steam. The steam-water mixture runs through a flash tank, which allows more of the water to change into steam. The plant separates the steam and directs it toward the turbine to generate electricity. Then the plant injects the remaining water back into the earth through an injection well. Once the steam cools in a condenser, it is also injected back into the ground.

> " The design of a geothermal power plant depends on whether the geothermal reservoir produces dry steam, wet steam, or hot water. "

To drive an electricity turbine, a power plant needs a great deal of high-temperature, high-pressure steam. A binary cycle plant or moderate-temperature system is used when a geothermal reservoir is not hot enough to produce significant amounts of steam. To solve this problem, the plant passes the water through a heat exchanger, where it heats a second liquid called a working fluid in a closed loop. Because the working fluid has a lower boiling point than water, it more easily creates a vapor at the lower temperature. The working fluid changes from liquid to vapor form and runs the turbine. It is then cooled and condensed back into a liquid and reused.

The binary cycle plant has several advantages. Using the working fluid, it can generate electricity at lower temperatures. This increases the number of reservoirs that are capable of generating electricity worldwide. In addition, the closed-loop system of a binary plant is environmentally friendly, with virtually no emissions.

Despite its advantages, the binary cycle plant is the least efficient geothermal plant design. To generate the same amount of electricity, it must circulate more water through the system as compared to a dry steam or flash steam plant. This higher flow rate and more complicated design can cause binary cycle plants to be more expensive than other geothermal power plants.

Can Geothermal Power Supply the World's Energy Needs?

As energy demand increases, many believe that geothermal energy may be part of a solution to supply the world's energy needs. Geothermal energy is renewable and reliable. Unlike solar or wind power, it is available 24 hours per day, which makes it well suited to provide baseline power needs. "If we can drill and recover just a fraction of the geothermal heat that exists, there will be enough to supply the entire planet with energy—energy that is clean and safe,"[3] said Are Lund, senior researcher at SINTEF Materials and Chemistry.

A major drawback to geothermal energy is that it is only available in certain areas. Current technology can access geothermal energy in places where hot magma rises close to the earth's surface. Outside of those areas, geothermal energy is limited. In the United States geothermal power plants are scattered across the western states, while people living in the East and Midwest have limited access to geothermal energy. "The big challenge is to show you can do it not only in California, but also in the Midwest and ultimately on the East Coast, where you have to go deeper,"[4] said Jefferson Tester, a professor of sustainable energy systems at Cornell University.

Can Geothermal Power Reduce Dependence on Fossil Fuels?

Today the world's demand for energy is primarily met by fossil fuels such as oil, coal, and natural gas. The supply of fossil fuels, however, is limited and nonrenewable. Most experts agree that as the fossil fuel supply

Hot magma sometimes reaches the earth's surface as lava and forms volcanoes but most of the time magma remains below the earth's crust, where it heats surrounding rocks and ground water. Hot lava pours from a red-hot volcano in this photograph.

dwindles, the world will need alternative energy sources.

Many believe that geothermal energy is well suited to replace fossil fuels for electricity generation and heating. "This is a very large resource that perhaps has been undervalued in terms of the impact it might have on supplying energy to the U.S.," said Tester. "Geothermal . . . could be much more compatible with our existing grid system than other renewables."[5] Available 24 hours per day, geothermal energy is a reliable energy source and can be used like fossil fuels to provide baseline power, which is the minimum amount of power needed by consumers.

> **Although reliable, geothermal power plants are costly to build.**

Although reliable, geothermal power plants are costly to build. High up-front costs to locate geothermal resources, drill, and build power plants are a barrier to use when compared to the less expensive fossil fuel power plants. However, if the price of fossil fuels continues to increase, geothermal power plants may become a cost-effective alternative energy source.

How Does Geothermal Power Affect the Environment?

Geothermal energy is one of the cleanest energy sources. It has lower emissions and pollution than fossil fuels. It uses less land than solar panels and wind farms and does not create harmful toxic waste like nuclear power plants. Geothermal energy is "a heck of a lot cleaner than burning coal or having to dispose of nuclear waste,"[6] said Ronald DiPippo, a mechanical engineer at the University of Massachusetts–Dartmouth.

Despite its clean benefits, geothermal energy has environmental drawbacks. Although emissions are lower, there is still the risk that bringing underground water to the surface will contaminate local water supplies and ground with naturally occurring but potentially toxic minerals and chemicals. In addition, the process of harnessing geothermal energy has been linked to earthquakes. Dennis Gilles, senior vice president for geothermal operations at Calpine Corporation in California, said, "We generate between 3,000 and 5,000 earthquakes a year. In a typical day, we experience on average 10." But Gilles said these quakes are very small;

they usually measure between 0.5 and 3.0 on the Richter scale. "I can guarantee you won't feel any of them,"[7] Gilles added.

However, a geothermal project in Basel, Switzerland, was halted in 2009 after it triggered an earthquake felt by nearby residents. "You are not going to want to put a geothermal facility like this where you have a danger of lubricating a big fault,"[8] said Tester.

What Is the Future of Geothermal Power?

In the coming years the world faces many uncertainties about energy supplies and sources. Many believe that geothermal energy has the potential to play a significant role in moving the world to a cleaner, more sustainable energy system.

With new technologies in development to drill deeper and draw geothermal energy from hot dry rock, the future of geothermal energy appears bright. "The fireball that sits within the Earth is a resource. We walk on it, we sleep on it, we work on it; the question is: How do we harness it?"[9] said Ólafur Ragnar Grímsson, president of Iceland, a country largely heated and powered by the earth's heat.

Can Geothermal Power Supply the World's Energy Needs?

"Most of the Earth below our feet is very hot. It's just a matter of knowing where to tap it."

—Mike Rhodes, professor of volcanology at the University of Massachusetts–Amherst.

"If world population and per capita use of energy continue growing at current rates or higher, our demand for energy is likely to grow faster than our ability to supply it from renewable sources."

—Ross McCluney, retired principal research scientist at the University of Central Florida.

World Energy Demand

Around the world, energy is a hot commodity. People use energy in homes, businesses, and factories. Energy powers the vehicles that transport goods and people and heats homes and buildings. According the Energy Information Administration, the world consumed 483.5 quadrillion Btu (British thermal unit) of energy in 2007. The United States was the largest consumer, using 101.5 quadrillion Btu, or approximately 21 percent of the world's total. After the United States, China, Russia, Japan, and India were also major energy users.

Currently, most energy consumed around the world comes from nonrenewable fossil fuels such as oil, natural gas, and coal. According to the Energy Information Association, 92 percent of energy consumed in the United States in 2009 was generated from nonrenewable fossil fuels. Although these fuels have been a cheap and convenient source of energy, they will not last forever. "The development of new and reliable energy sources is vital to sustain our way of life—it is currently one of the most important challenges the international community faces,"[10] said Mike Barnes, geothermal team leader for Mighty River Power, a geothermal company in New Zealand.

As a result, there has been much interest in developing renewable energy sources such as geothermal energy, which can be replenished. According to Leslie Blodgett, outreach director of the Geothermal Energy Association, "Recognition of geothermal energy's value as a zero-emissions base-load energy source is growing."[11]

Geothermal Power Production Worldwide

Geothermal energy is already being used in several countries to generate electricity. In 2010 geothermal power stations were forecasted to have a total installed capacity of more than 12,000 megawatts. According to the Geothermal Energy Association, this amount of energy could supply power to more than 52.5 million people in 24 countries.

The United States is the world's leader in geothermal power generation. In 2009 the United States generated 15.2 billion kilowatt-hours of geothermal power, which was 0.4 percent of total US electricity generation. The majority of geothermal plants in the United States are located in California, which produced almost 86 percent of the United States' geothermal electricity in 2009. In addition, geothermal power plants operate in several western states, including Nevada, Hawaii, Idaho, and Utah. Outside the United States, several countries, including the Philippines, Indonesia, Guatemala, Australia, and Costa Rica, use geothermal energy to generate electricity.

The World's Largest Geothermal System

One of the United States' most productive sources of geothermal power lies in an area north of San Francisco, California, called the Geysers. The Geysers, which celebrated its fiftieth anniversary in 2010, is the world's largest geothermal energy operation. Eighteen power plants produce 850 megawatts of electricity annually, which is enough to power about

850,000 homes. "It generates enough clean electricity to power a city the size of San Francisco. The Geysers serves as a thriving example of innovation and part of the answer to our need for resources to power the world we live in,"[12] said Mike Rogers, a senior vice president at Calpine Corporation, a company that owns and operates several power plants at the Geysers.

The Geysers sit on top of a dry steam geothermal reservoir. About 4 miles (6.4km) beneath the surface, a magma formation that has split the earth's crust heats underground rocks. When water seeps below the surface and meets the heated rocks, it turns to steam. Steam production wells can stretch more than 2 miles (3.2km) deep to reach this superheated, dry steam. When the steam shoots through the pipes to the surface, it flows through a network of connected power plants, where it spins turbines to generate electricity.

> **The United States is the world's leader in geothermal power generation.**

Over the years rapid expansion at the Geysers led to a decrease in steam output. With less steam to run turbines, the power plants produced less electricity. Some scientists feared that the Geysers would eventually run out of steam because it had been developed too quickly. In 1997, however, plant operators learned to create more steam by injecting water back into the Geysers' underground reservoir. "Getting water back in the ground to recharge the reservoir and using steam more efficiently is the key. . . . We've learned a huge amount about how to better utilize the resource. The Geysers is going to be around for another 50 years,"[13] said Rogers.

Geothermal Energy in Homes and Businesses

While high-temperature geothermal resources can produce electricity, low- to moderate-temperature reservoirs can be used in a variety of direct-use applications. District heating systems pipe hot water near the earth's surface through a series of pipes to houses and buildings to provide heat. For more than a century, the towns of Klamath Falls, Oregon, and Boise, Idaho, have heated homes and businesses with direct-piped geothermal water. Additionally, some cities use direct-piped hot water under roads and sidewalks to melt ice and snow during the winter.

In Midland, South Dakota, hot water from a geothermal spring heats the town's school. A 3,300-foot well (1,006m) circulates hot water from the nearby Yellowstone aquifer, a porous layer of underground rock that holds water, to keep the building warm. The hot spring water flows through geothermal water pipes to radiated heaters in 27 classrooms. "It gets pumped through the six inch line and it goes all the way out throughout the building to each individual classroom. . . . The heat from that water just heats up and then it's got the fans that turn on and off with the thermostat settings and it just blows in the warm air into the building,"[14] said school maintenance worker Mikel Williamson.

Elsewhere, hot spring water also heats pools for bathing and swimming. Japan has more than 2,800 spas, 5,500 public bathhouses, and 15,600 hotels that use geothermal hot water. Geothermal energy heats swimming pools in Iceland and Hungary, many of which are year-round, open-air pools. At the Glenwood Hot Springs in Glenwood, Colorado, guests can swim in one of the world's largest outdoor geothermal pools. The 450-foot pool (137m) uses hot spring water that rises through fractured rock at 122°F (50°C) before being cooled down to a temperature ranging from 90°F to 104°F (32°C to 40°C).

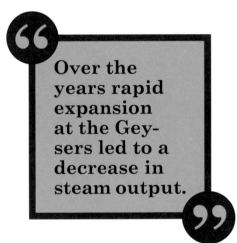

Over the years rapid expansion at the Geysers led to a decrease in steam output.

Greenhouses and aquaculture (fish farming) also use direct geothermal energy. During cold months geothermal heat warms greenhouses to grow fresh vegetables and plants in Russia, Hungary, Iceland, and the United States. Tuscany, a region in central Italy, grows vegetables in the winter from fields heated by natural steam. In California, fish farms use warm water from the earth to produce approximately 10 million pounds (4.5 million kg) of tilapia, striped bass, and catfish. Industrial applications use geothermal water for food dehydration, gold mining, and milk pasteurization.

In addition to direct-piped geothermal applications, geothermal heat pumps replace energy needed to heat and cool homes and businesses.

The heat pump works because only a few feet below the surface, the earth remains a relatively constant temperature year-round. The heat pump transfers heat from the earth into a building during the winter months and transfers it from the building into the ground during warmer summer months. According to the US Department of Energy, approximately 50,000 geothermal heat pumps are installed annually. "That's a lot of usage there, but it's still small compared with the overall U.S. heating and cooling market,"[15] said Cliff Chen, senior energy analyst for the Union of Concerned Scientists, a science advocacy group.

Limited Access to Geothermal Resources

Despite geothermal energy's potential to generate electricity and heat homes and businesses, it is generally limited to regions near underground geothermal reservoirs. These reservoirs are usually found near major continental plate boundaries, where earthquakes and volcanoes are common. Current geothermal technologies can reach the earth's thermal energy in these places, because the hot magma is close to the surface and heats ground water to temperatures above 212°F (100°C). These prime geothermal locations are concentrated in specific areas of the world, such as Iceland, the Philippines, and countries along the Ring of Fire in the Pacific Ocean. Outside of these areas, geothermal energy is often not available.

> **Even in areas with promising geothermal conditions, economically usable sites are difficult to find.**

Although the United States is the world's leading geothermal power producer, existing power plants are concentrated in the western states. While these plants can provide power for the West, they are too far away to be an economical power solution for eastern population centers. For the rest of the country, the challenge is to access the geothermal energy found in hot dry rock without an existing underground water or steam reservoir. "Hot rock is present across the United States, but new methods have to be developed to use the heat in these rocks to produce geothermal power,"[16] said Joe Moore, a geologist with the University of Utah's Energy & Geoscience Institute.

Usable Sites Hard to Find

Even in areas with promising geothermal conditions, economically usable sites are difficult to find. Lucien Bronicki, chair and chief technology officer of Ormat Technologies, a geothermal company, explained:

> We need three elements, really, to have a good resource. One is to have heat. This is relatively easy to detect. But you need two other things. You need water, which is the element that brings the heat from the depths to the surface. And you need rocks which are the edge of fractures, a fault or a permeable, so that this water can flow, so that when you drill a well the water tends to flow into the well and come out. . . . And this exploration takes time.[17]

Additionally, a geothermal plant needs to be near a source of surface water for cooling equipment used in generating electricity. Given that many geothermal resources are concentrated in dry, arid regions, an adequate surface water source can be difficult to find. Additionally, geothermal sites need to be near power transmission lines so that electricity can be transferred to users.

While many energy researchers believe that geothermal is a promising energy source, the process of finding suitable sites for generating an adequate supply of electricity is fraught with unknowns, and the cost of such exploration is high. "You might put $20 [million] to $40 million into digging holes just to find out what the capacity of your field is. No one will sign contracts until you know for sure,"[18] said Brett King, whose law firm has been involved in several geothermal deals.

> " Scientists are working to find ways to better identify and evaluate geothermal resources. "

Expanding Access to Geothermal Energy

Although worldwide access to geothermal energy is limited, new research to expand usable sites appears promising. Scientists are working to find ways to better identify and evaluate geothermal resources. "The United

States Geological Survey estimates that 70 to 80 percent of U.S. geothermal resources are hidden," said Karl Gawell, executive director of the Geothermal Energy Association. "You can't see it on the surface, and we don't have the technology to find it without blind drilling. . . . Geothermal hasn't had the breakthroughs in geophysical science that the oil industry had in 1920s. We are still looking for where it's leaking out of the ground."[19]

In addition, researchers are working on ways to drill to deeper depths to reach hotter temperatures. They also hope to unlock geothermal energy in hot dry rock formations that are more common around the world. These improvements may lead to more universal access to geothermal energy worldwide.

Vast Potential for Geothermal Energy

Geothermal energy holds vast potential to provide electricity and heat homes around the world. Many countries that border the Pacific Ocean or Africa's Great Rift Valley are rich in geothermal energy resources. Other areas have the ability to use lower-temperature geothermal resources for direct-use applications.

Currently, drilling technology limits access to geothermal energy. With advances in drilling technology, this barrier might eventually be lifted. In 2006 a group of scientists and engineers met at the Massachusetts Institute of Technology to assess the United States' geothermal electrical generating potential. They concluded that if the technology to access thermal energy in hot dry rock could be developed, the resulting geothermal energy would provide more than 2,000 times the United States' energy needs. While acknowledging that the technology to use geothermal energy on a widespread basis does not currently exist, scientists and engineers at the Massachusetts Institute of Technology affirm its potential. They write: "It is important to emphasize that while further advances are needed, none of the known technical and economic barriers limiting widespread development of [geothermal energy] as a domestic energy source are considered to be insurmountable."[20]

Can Geothermal Power Supply the World's Energy Needs?

66 Geothermal energy . . . is capable of providing enormous supplies of electricity for America. 99

—Al Gore, "Al Gore's Speech at Constitution Hall in Washington," National Public Radio, July 17, 2008. www.npr.org.

Gore is a former vice president of the United States.

66 If geothermal is going to be anything more than a minor curiosity, it has to reach at least the level of hydro and nuclear power, or 100,000 megawatts out of 1 million—one-tenth of total capacity. 99

— Jefferson Tester, "Chemical Engineer: Geothermal Is Undervalued US Energy Source," Massachusetts Institute of Technology, February 13, 2007. http://web.mit.edu.

Tester is the former H.P. Meissner Professor of Chemical Engineering at the Massachusetts Institute of Technology.

* Editor's Note: While the definition of a primary source can be narrowly or broadly defined, for the purposes of Compact Research, a primary source consists of: 1) results of original research presented by an organization or researcher; 2) eyewitness accounts of events, personal experience, or work experience; 3) first-person editorials offering pundits' opinions; 4) government officials presenting political plans and/or policies; 5) representatives of organizations presenting testimony or policy.

66 **There is the potential to power millions of homes, businesses, and schools from the heat of the earth.** 99

—Karl Gawell, "Geothermal's Golden Year: After 50 Years Geothermal Energy Still Growing," Geothermal Energy Association, August 23, 2010. http://geo-energy.org.

Gawell is the executive director of the Geothermal Energy Association.

66 **Hot rock is present across the United States, but new methods have to be developed to use the heat in these rocks to produce geothermal power.** 99

—Joe Moore, "Making Geothermal More Productive," University of Utah, September 8, 2009. www.unews.utah.edu.

Moore is a geologist at the Energy & Geoscience Institute at the University of Utah.

66 **The real need for innovation and development needs to come from the distribution side . . . if we can't get the power from where it is generated to where it is needed, it's just an exercise in expensive futility.** 99

—Sterling Burnett, "Q&A: Experts Debate the Market Outlook for Renewable Energy and Other 'Green' Initiatives," *Dallas Business Journal*, October 25, 2009. www.bizjournals.com.

Burnett is a senior fellow and analyst at the National Center for Policy Analysis in Dallas.

66 **Mother Nature has bestowed upon us the abundant geothermal resources. It is now our duty to develop these resources.** 99

—Kiraitu Murungi, speech at the East African Rift System Regional Stakeholders' Workshop, March 15, 2010. www.africa-union.org.

Murungi is Kenya's minister for energy.

66 There are significant obstacles today to scaling up geothermal to serve a global customer base because most of the current investment in geothermal is going toward localized development of shallow and known geothermal resource areas that tend to be limited in size. 99

—Daniel Kunz, "Beyond Fossil Fuels: Daniel Kunz on Geothermal Energy," *Scientific American*, April 24, 2009. www.scientificamerican.com.

Kunz is president and chief executive officer of US Geothermal.

66 It is my strong hope that the efforts to tap the potential of geothermal energy will be successful, not only in Indonesia but also everywhere else in the world, where geothermal energy reserves are still to be made serviceable. 99

—Susilo Bambang Yudhoyono, speech at the World Geothermal Congress, April 26, 2010. http://indonesia.gr.

Yudhoyono is president of the Republic of Indonesia.

Facts and Illustrations

Can Geothermal Power Supply the World's Energy Needs?

- The heat continuously flowing from the earth's interior is estimated to be equivalent to **42 million megawatts** of power and is expected to remain so for billions of years.

- Domestic geothermal resources at a depth of 1.8 miles (3km) are estimated to be about 3 million quads (one quad equals 170 million barrels of oil), which is equivalent to a **30,000-year energy supply** at the United States' current energy consumption rate.

- Annual US energy consumption equals about 100 quads, the equivalent of 17 billion barrels of oil per year. In 2009 geothermal power contributed around **2 percent** to the US energy mix.

- Heat pumps run by geothermal energy are used by households and businesses in all **50 states**.

- Worldwide there are more than **half a million** geothermal heat pumps installed, for a total thermal output of over 7,000 megawatts.

- In 2009 geothermal energy was the **sixth-largest source** of renewable energy in the United States, behind hydropower, wood, biofuels, wind, and biomass waste.

Binary Cycle Power Plants

To increase access to geothermal energy, binary cycle plants can be used to capture geothermal energy in areas with lower temperature geothermal reservoirs. A binary cycle plant heats a second liquid called a working fluid, which has a lower boiling point than water and more easily creates a vapor at lower temperatures. Because binary cycle plants run on a closed loop, they emit virtually no emissions into the atmosphere.

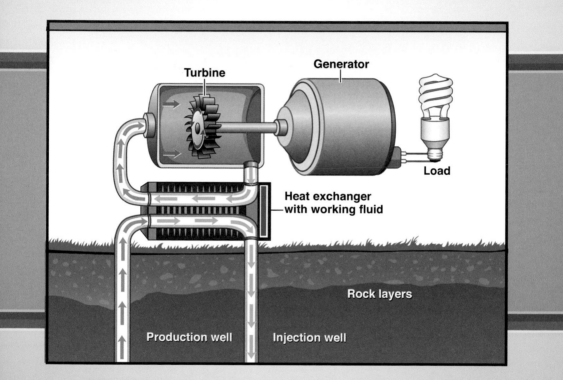

Source: US Department of Energy, "Hydrothermal Power Systems," November 2010. www.eere.energy.gov.

- Geothermal energy currently supplies more than 12,000 megawatts to 24 countries worldwide and produces enough electricity to meet the needs of **60 million people**.

- As of 2010, **California** has more geothermal power **online** than any country with geothermal power.

US Geothermal Resources by State

Existing and planned geothermal resources are centered in western states where geothermal reservoirs have been found. In order for geothermal energy to be more widely used, scientists are researching new technologies and drilling methods that will extend the available geothermal resource to more areas of the country.

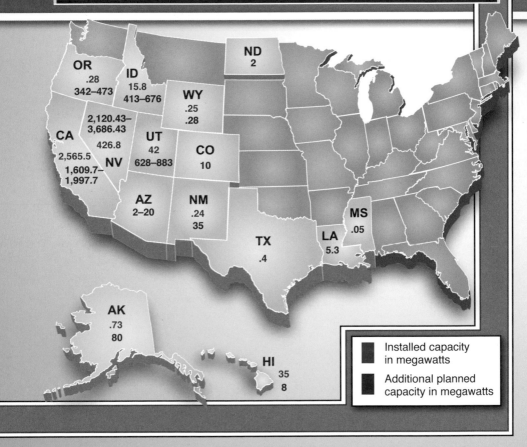

Source: National Renewable Energy Laboratory, "Geothermal Power Generation," May 2010. www.nrel.gov.

- According to the Earth Policy Institute, more than **150 power plants** were under development in 13 US states as of 2010; these plants are expected to almost triple geothermal generating capacity in the United States.

Hydrothermal Reservoir

Currently, access to geothermal energy is limited to areas of the world that are located near underground geothermal reservoirs. A geothermal reservoir operates as a hydrothermal system that circulates fluids through hot rock and transfers heat energy upward. The fluids in a hydrothermic system can be steam vapor or hot water. Rain and surface water seep through cracks and permeable rock to the underground reservoir where it is heated by hot rock.

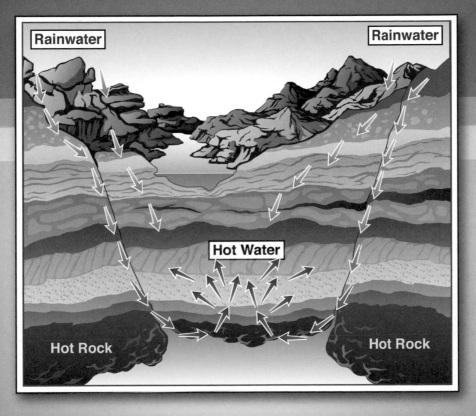

Source: Bruce D. Green and R. Gerald Nix, "Geothermal—The Energy Under Our Feet," National Renewable Energy Laboratory, November 2006. www.eere.energy.gov.

- As of 2010, **24 countries** use geothermal energy to generate electricity.

Geothermal Energy by Country

Although the United States is the world leader in geothermal energy production, many other countries are using energy from the earth. As of 2010, development of geothermal energy is occurring on six continents.

Estimated Production of Geothermal Energy by Country and Megwatts in 2010

Source: US Department of Energy, "Geothermal Tomorrow: 2008," September 2008. www.nrel.gov.

- The most common non-electric uses worldwide of geothermal energy are **heat pumps, bathing, space-heating, greenhouses, aquaculture, and industrial processes**.

Can Geothermal Power Reduce Dependence on Fossil Fuels?

66Deep geothermal energy is a real hot prospect as we dig deeper for new technologies that cut carbon emissions and provide home-grown power.99

—Chris Huhne, UK energy and climate change secretary.

66There is no way America can end its addiction to fossil fuels. Solar, geothermal, and wind power just aren't there yet. . . . In 15 years—by even the most optimistic forecast—we will still have 75% of our energy coming from fossil fuels.99

—Christian DeHaemer, editor of *Wealth Daily*.

Dependence on Fossil Fuels

For the past century fossil fuels such as oil, coal, and natural gas have been economical and accessible energy sources. Petroleum fuels cars, truck, and airplanes and is used to manufacture plastics. Coal and natural gas heat homes and generate electricity to run lights and everyday appliances such as computers, refrigerators, and air conditioners.

According to the International Energy Association, nearly 80 percent of the energy consumed worldwide in 2008 came from fossil fuels. In addition, fossil fuels generated approximately 67 percent of the world's electricity. In contrast, alternate energy sources such as geothermal, solar, and wind energy contributed about 3 percent of world energy needs and less than 3 percent of world electricity generation.

The United States is one of the world's largest consumers of fossil fuels. According to the Energy Information Administration, oil, coal, and natural gas provided 83 percent of energy consumed in 2009. Renewable energy sources, including geothermal energy, provided only 8 percent of energy consumed in 2009.

Concerns About Fossil Fuels

Although fossil fuels have historically been an easy-to-use and reliable energy source, there are several concerns about relying on them for future energy needs. First, fossil fuels are not renewable. Some experts believe that the world has already reached peak production for fossil fuels or will reach it shortly. As fossil fuel reserves become depleted, they will become more expensive. As a result, it will cost more to drive cars, heat homes, and purchase products manufactured with oil. "Something has to give, and what will give is the price of energy. It's got to go up,"[21] said Stephen Mayfield, director of the San Diego Center for Algae Biotechnology.

According to the International Energy Association, nearly 80 percent of the energy consumed worldwide in 2008 came from fossil fuels.

In addition, the United States imports more than half of its oil from foreign countries. According to Ray Mabus, the secretary of the US Navy and a former ambassador to Saudi Arabia, fossil fuels often come from unstable regions of the world, where dwindling supplies can be a cause of international conflict. Developing homegrown, renewable sources of energy such as geothermal could potentially improve national security.

Another concern is fossil fuels' effect on the environment. Burning fossil fuels releases pollutants into the air that have been linked to acid rain, greenhouse gases, and human health impacts. In addition, the process of extracting fossil fuels from the earth can damage ecosystems and habitats. In a 2010 speech supporting alternative energy research, President Barack Obama said that continued fossil fuel dependency "will jeopardize our national security, it will smother our planet and will continue to put our economy, and our environment at risk. . . . The time has come, once and for all, for this nation to fully embrace a clean energy future."[22] The concerns with fossil fuels have led to an increased focus on alternative energy sources such as geothermal energy.

A Reliable Energy Source

Supporters of geothermal energy believe it is a suitable alternative to fossil fuels. Geothermal plants are simple in design, so there is less to go wrong as compared to other types of power plants. Geothermal power is also an extremely reliable energy source. With its fuel source directly beneath it, the geothermal plant can operate without regard to weather, natural disasters, or political conflicts. This is an advantage over fossil fuel power plants, which depend on the transport of fuel for operation; any of these events can disrupt fuel transports to the power plant.

One of the primary benefits of geothermal energy is its ability to provide base-load power. Base-load power plants supply power for most of day, in contrast to peak plants, which provide intermittent power during peak demand times. According to a report by the National Geothermal Collaborative, geothermal power plants are available to operate approximately 98 percent of the time. This compares favorably to nuclear and fossil fuel power plants, which run between 75 to 90 percent of the time. "Geothermal is an extremely reliable form of energy, and it generates power 24/7, which makes it a base-load source like coal or nuclear,"[23] said David Blackwell, director of the Southern Methodist University Geothermal Laboratory. In addition to providing base-load power needs, geothermal power plants can be installed in incremental units as electricity demand increases.

> A typical geothermal well costs between \$1 million and \$5 million.

Among alternative energy sources, geothermal has the ability to provide full-time power, regardless of the weather conditions or time of day. In contrast, wind and solar power facilities generate power intermittently, with unavoidable down time. "Wind doesn't always blow and the sun doesn't always shine,"[24] said Daniel Kunz, a representative from Boise-based US Geothermal. Using geothermal for base-load power would eliminate power generation down time.

Is Geothermal Energy Truly Renewable?

The Energy Department classifies geothermal energy as a renewable resource. The earth's core provides an almost unlimited amount of heat. Yearly rainfall and melting snow constantly supply new water to geothermal reservoirs. Production from geothermal fields can be sustained for decades and longer. The Larderello, Italy, power plants that have been operating since 1913 and the Geysers in California, which began operation in 1960, are working examples of the long-term sustainability of geothermal energy.

Some experts point out, however, that geothermal energy is not truly renewable. Although the earth's heat is constant, the geothermal reservoir may be damaged or depleted if a power plant withdraws too much steam and hot water too quickly without giving the source a chance to replenish itself.

Although the initial costs may be high, geothermal power plants cost less to operate and maintain than fossil fuel power plants.

When the Geysers initially began operations, it started out slowly but then quickly expanded in the 1970s and 1980s with additional wells extracting steam. The Geysers' steam production peaked in the late 1980s and then rapidly declined after too many wells pulled from the same steam resource. In the process of searching for a solution to this problem, engineers discovered that injecting additional water into the steam field helped to stabilize the Geysers' output. They believe that this solution will allow them to sustain the Geysers for decades.

According to John Farison, an engineer for Calpine Corporation at the Geysers, the key challenge to sustaining a geothermal field is to manage the steam resource correctly. If engineers inject too much water into an area, it can cool the geothermal reservoir and hurt production. To avoid this, Farison monitors the output in adjacent steam wells. An increase or decrease in steam production tells him if he is injecting too much or too little water into his wells. He says that by carefully managing where and how much water is reinjected into the reservoir, engineers can optimize steam output.

The High Cost of Geothermal Power

Developing a geothermal resource is a complicated, expensive process. According to the National Geothermal Collaborative, only about 1 in 5 exploratory wells drilled confirms a valuable and usable geothermal resource. Costs depend on the location, temperature, and drilling depth of wells. A typical geothermal well costs between $1 million and $5 million. The least expensive wells are generally those that produce high temperatures at shallow depths. Even in relatively shallow wells, however, drilling costs can run from one-third to one-half of the total project's cost. In addition, because many geothermal resources are located in remote areas, communities incur additional expenses to build power generation lines to transmit the electricity to users.

To date, the higher up-front costs to develop geothermal energy have made it a less attractive option than fossil fuels and other alternative energy sources. According to the Energy Information Administration, approximately 76 percent of the United States' electricity generating capacity came from fossil fuel–fired power plants in 2008. In contrast, more expensive geothermal power plants have been slower to develop, contributing less than 1 percent of electricity generating capacity in 2008.

Direct-use applications of geothermal energy also have high up-front costs. According to Guy Marshall, general manager of a company that installs residential geothermal systems, a closed-loop geothermal forced-air system to heat and cool a home may cost anywhere from $15,000 to $40,000, depending on the size of the home.

> " Dollars spent on geothermal energy stay in local economies and reduce the country's dependence on imported energy. "

Although the initial costs for geothermal systems are high, drilling costs have dropped approximately 25 percent since the early 1990s because of technological advances. As drilling technology continues to improve, costs should keep dropping. "We cannot be so concerned about the initial cost. Ultimately, the cost will go down, the technology will improve,"[25] said Gary Locke, the US secretary of commerce.

Hidden Costs of Fossil Fuels

Some scientists believe simply comparing the costs to build a fossil fuel power plant to a geothermal power plant does not provide an accurate picture. They argue that fossil fuel power plants incur additional environmental costs that should be considered in the analysis. These costs include human health problems caused by harmful air emissions, destruction of plants and animals when drilling or mining for fossil fuels, and environmental damage from global warming, acid rain, water pollution, and land degradation. According to the Union of Concerned Scientists, "These costs are indirect and difficult to determine, they have traditionally remained external to the energy pricing system . . . , but this pricing system masks the true costs of fossil fuels and results in damage to human health, the environment, and the economy."[26]

According to the Geothermal Energy Association, the cost of coal-fired power plants would increase by 25 percent and natural gas plants by 17 percent if environmental costs were included. This brings geothermal energy systems closer in cost to fossil fuel systems.

Lower Operating Costs

Although the initial costs may be high, geothermal power plants cost less to operate and maintain than fossil fuel power plants. Unlike a geothermal plant, which sits on top of free thermal energy, fossil fuel plants incur fuel costs to generate electricity for the life of the plant. Over a 30-year period, fuel costs for natural gas and coal plants can be significant, accumulating to twice the initial plant capital costs. In addition, a geothermal plant incurs no fuel transportation costs. Factoring in lower operating and maintenance costs makes geothermal energy less expensive. "It has the lowest levelized cost of any power source in the world, even coal,"[27] said Alison Thompson, chair of the Canadian Geothermal Energy Association.

Direct-use geothermal systems also provide operation and maintenance savings to consumers. In 2008 Robert Hayward replaced his New Hampshire home's oil-heating system with a geothermal system. Although the system installation cost $30,000, he is already recovering the expense through lower heating and cooling bills. "We are able to save about half on our energy costs. About 50–60 percent. That's a savings of about $2,000 per year on our energy costs,"[28] said Hayward.

Unlike fossil fuels, geothermal energy reduces the United States' reliance on foreign governments for energy. Domestically produced geothermal energy is not susceptible to fuel price shocks like the United States experienced during the 1973 oil embargo by the Organization of the Petroleum Exporting Countries, which caused gas prices to skyrocket.

In contrast to imported fossil fuels, geothermal energy is generated with local resources. Dollars spent on geothermal energy stay in local economies and reduce the country's dependence on imported energy. Geothermal plants in Nevada that produce about 210 megawatts of electricity save energy imports that are equal to 800,000 tons (725,748 metric tons) of coal or 3 million barrels of oil each year.

In addition, because geothermal energy is locally produced, local and state governments benefit from tax revenue and job creation. The US Bureau of Land Management also collects millions of dollars each year in rent and royalties from geothermal power plants on federal lands. "It's clean. It's local. It's here—it's indigenous,"[29] said Michael Kaleikini, plant manager for Puna Geothermal Venture in Hawaii.

Replacing Fossil Fuels

Today the world's energy demand is primarily met by fossil fuels such as oil, coal, and natural gas. The supply of fossil fuels, however, is limited and nonrenewable. Most experts agree that the world needs to look for alternative energy sources like geothermal energy.

High up-front costs, however, have been a barrier for geothermal development. Because fossil fuel power plants have been cheaper and more convenient, they have been built at a faster rate. However, if the price of fossil fuels increases as the supply declines, geothermal power plants may become a comparable or even more cost-effective energy source. "There are thousands of wells being drilled for oil across the world every year. I imagine that in a couple of decades all of those drilling rigs that are now redundant, because we've run out of oil, will be drilling geothermal wells instead,"[30] said Doone Wyborn, a scientist with Geodynamics, an Australian geothermal company.

Primary Source Quotes*

Can Geothermal Power Reduce Dependence on Fossil Fuels?

"As we get serious about addressing global warming, geothermal will grow in importance because it can more effectively replace conventional coal-fired power plants."

—Karl Gawell, "Geothermal's Golden Year: After 50 Years Geothermal Energy Still Growing," Geothermal Energy Association, August 23, 2010. http://geo-energy.org.

Gawell is the executive director of the Geothermal Energy Association.

"The transition to a lower-carbon economy won't happen overnight. The sheer scale of the energy industry makes this impossible. . . . Long lead times mean that like it or not, fossil fuels will continue to play a very significant part in the future energy mix."

—Tony Hayward, "Meeting the Energy Challenge," speech at the Oil and Money Conference, London, October 20, 2009. www.bp.com.

Hayward is the former chief executive officer of oil and energy company BP.

* Editor's Note: While the definition of a primary source can be narrowly or broadly defined, for the purposes of Compact Research, a primary source consists of: 1) results of original research presented by an organization or researcher; 2) eyewitness accounts of events, personal experience, or work experience; 3) first-person editorials offering pundits' opinions; 4) government officials presenting political plans and/or policies; 5) representatives of organizations presenting testimony or policy.

Primary Source Quotes

> ❝As the only baseload source of renewable power, geothermal is the perfect option for utilities interested in green power.❞

—Karl Gawell, "Geothermal Energy Association Workshop to Address Utilities, Co-op, and Public Power," Geothermal Energy Association, June 17, 2010. http://geo-energy.org.

Gawell is the executive director of the Geothermal Energy Association.

....

> ❝A significant amount of the financial risk associated with geothermal power development results from uncertainties encountered in the early stages of resource development; namely geothermal exploration and the drilling of production wells.❞

—Dan Jennejohn, *Research and Development in Geothermal Exploration and Drilling*, Geothermal Energy Association, December 2009. http://geo-energy.org.

Jennejohn is a research associate with the Geothermal Energy Association.

....

> ❝Geothermal wells are like oil wells—some wells produce and some don't. Drilling wells is expensive. That is why we need to develop low-cost techniques to improve their productivity.❞

—Joe Moore, "Making Geothermal More Productive," University of Utah, September 8, 2009. www.unews.utah.edu.

Moore is a geologist at the Energy & Geoscience Institute at the University of Utah.

....

> ❝There's a significant risk to developing geothermal: high upfront costs of drilling and establishing your wells. But then once they're drilled, it's just a matter of maintaining them. Your going forward cost is competitive in the energy market.❞

—Larry Sessions, "Personnel Profile: Q&A with Larry Sessions," *Capitol Weekly*, November 18, 2010. www.capitolweekly.net.

Sessions is the general manager of the Geysers.

....

66 Geothermal power plants are cost effective, reliable, sustainable, and environmentally friendly. The adoption of this technology is key to achieving our country's sustainability goals. 99

—J.T. Grumski, "SAIC Awarded $14 Million Contract to Deliver Renewable Energy Generation Facility," press release, PR Newswire, October 5, 2010. www.prnewswire.com.

Grumski is senior vice president and business unit general manager for SAIC, a scientific, engineering, and technology applications company.

66 With geothermal, you are spending a lot of money digging holes in the ground hoping to find these pockets of hot water. So you're just taking a lot of risk on the front end in making sure the resource is available and acceptable, and that's been difficult. 99

—Daniel Mannes, "Geothermal Energy Production and the California Election Provides Volatile Opportunity to Play the Future of Carbon Free Electricity," Wall Street Transcript, November 1, 2010. www.twst.com.

Mannes is a senior research analyst who covers the alternative energy sector for Avondale Partners.

66 To those who say the costs are still too high: I ask them to consider whether the costs of oil and coal will ever stop increasing if we keep relying on quickly depleting energy sources to feed a rapidly growing demand all around the world. 99

—Al Gore, "Al Gore's Speech at Constitution Hall in Washington," National Public Radio, July 17, 2008. www.npr.org.

Gore is a former vice president of the United States.

66 I think of geothermal power as the Cinderella of alternative energy, most people don't pay it any attention. It's really an attractive alternative. Unlike wind and solar energy, it works 24 hours a day, seven days a week, is unobtrusive and environmentally friendly. 99

—Mike Rhodes, "UMass Amherst and Connecticut Geologists Position New England for Success in the Geothermal Power Era," University of Massachusetts–Amherst press release, November 8, 2010. www.umass.edu.

Rhodes is a professor of volcanology at the University of Massachusetts–Amherst.

Can Geothermal Power Reduce Dependence on Fossil Fuels?

- Dependence on foreign countries is a concern with fossil fuels: The United States imports close to **60 percent** of its oil and nearly **10 percent** of its natural gas.

- In 2009 the United States consumed approximately **83 percent** of its energy needs from fossil fuels: coal, oil, and natural gas.

- If only **1 percent** of the thermal energy contained within 6.2 miles (10km) of the earth's surface were captured, this amount would be 500 times the energy contained in all oil and gas resources of the world.

- At current consumption rates, **1 megawatt-electric** is sufficient to supply a community with a population of 1,000.

- Initial construction costs of a geothermal facility represent **two-thirds** or more of total costs.

- The cost of coal-fired power plants would increase by **25 percent** and natural gas plants by **17 percent** if environmental costs were included.

- The price of geothermal energy **does not fluctuate** like the price of oil and gas.

Geothermal Power Plant Development Costs

High, up-front construction costs has limited the number of geothermal power plants built, as compared to cheaper fossil fuel power plants. Plant construction is the highest expense, followed by exploratory and production drilling. After the geothermal plant is operational, it has low operating and maintenance costs.

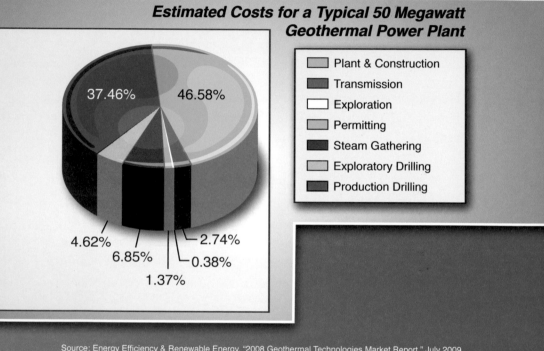

Estimated Costs for a Typical 50 Megawatt Geothermal Power Plant

- Plant & Construction
- Transmission
- Exploration
- Permitting
- Steam Gathering
- Exploratory Drilling
- Production Drilling

37.46% 46.58%

4.62% 2.74%
6.85% 0.38%
1.37%

Source: Energy Efficiency & Renewable Energy, "2008 Geothermal Technologies Market Report," July 2009. www.eere.energy.gov.

- A closed-loop geothermal forced-air system to heat and cool a home may cost anywhere from **$15,000 to $40,000**, depending on the size of the home, making geothermal systems more expensive that comparable fossil fuel heating and cooling systems.

- Geothermal energy's worldwide available resources have been calculated to be **larger** than the resource bases of coal, oil, gas, and uranium combined.

Fossil Fuels Dominate US Energy Picture

Most of the energy consumed in the United States comes from fossil fuels—oil (or petroleum), coal, and natural gas. Renewable energy sources such as solar and geothermal, have been more expensive to produce and use and have been limited by technology, availability, and location. In recent years, the production of renewable fuels has grown more quickly. As the prices of oil and gas increase, renewables are poised to play an important part in the US—and world—energy supply.

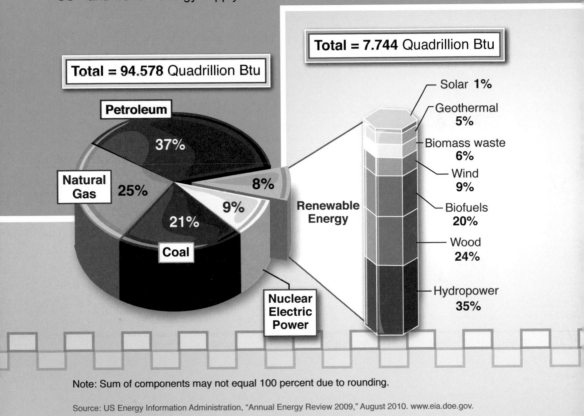

Total = 7.744 Quadrillion Btu

Total = 94.578 Quadrillion Btu

Petroleum 37%

Natural Gas 25%

Coal 21%

9%

8%

Renewable Energy

Nuclear Electric Power

Solar 1%
Geothermal 5%
Biomass waste 6%
Wind 9%
Biofuels 20%
Wood 24%
Hydropower 35%

Note: Sum of components may not equal 100 percent due to rounding.

Source: US Energy Information Administration, "Annual Energy Review 2009," August 2010. www.eia.doe.gov.

- Geothermal energy is considered a **renewable energy source** because the water is replenished by rainfall and the heat is continuously produced by the earth.

Other Ways to Use Geothermal Energy

In addition to replacing fossil fuels to produce electricity, geothermal energy is used in many direct-use applications around the world. The most popular direct-use is geothermal heat pumps, which can be used almost anywhere in the world because they do not require access to an underground reservoir of steam or hot water. Other direct uses include bathing, heating, and food dehydration.

Direct-use application breakdown by installed capacity and annual energy use

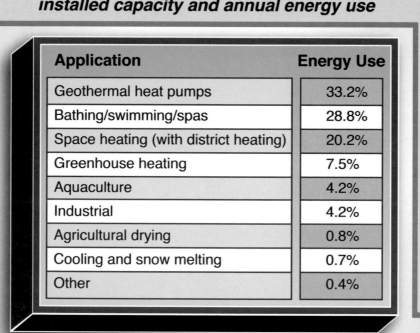

Application	Energy Use
Geothermal heat pumps	33.2%
Bathing/swimming/spas	28.8%
Space heating (with district heating)	20.2%
Greenhouse heating	7.5%
Aquaculture	4.2%
Industrial	4.2%
Agricultural drying	0.8%
Cooling and snow melting	0.7%
Other	0.4%

Source: Energy Efficiency & Renewable Energy, "2008 Geothermal Technologies Market Report," July 2009. www.eere.energy.gov.

- Although considered renewable, geothermal wells may stop producing energy over time because the normal **life span** for a deep geothermal well is around 30 years, after which the rock has cooled down too much from the cold water injected into the well.

How Does Geothermal Power Affect the Environment?

66We believe the thermal process to be a fairly small footprint, kind of unobtrusive and out in the middle of nowhere so no one is going to see it.99

—Larry Sessions, general manager of the Geysers.

66As long as you do this far away from inhabited areas, there shouldn't be a problem. But for cities with a history of earthquakes, it's probably best not to install enhanced geothermal.99

—Domenico Giardini, director of the Swiss Seismological Service.

Since 1930 Reykjavík, Iceland, has been drawing up hot water from underground springs to heat its buildings. In addition, Reykjavík uses geothermal energy to generate electricity. Using geothermal energy for heat and electricity has significantly reduced Reykjavík's dependence on fossil fuels. In addition, the switch to geothermal energy has benefited the environment, reducing the city's carbon dioxide emissions by approximately 4 million tons (3.6 million metric tons) annually. Before geothermal energy, smoke from burning coal filled the sky above Reykjavík in the 1920s. Today the same skyline has virtually no visible air pollution and is now one of the cleanest cities in the world.

Fossil Fuels Are Harmful to the Environment

Although fossil fuels have been a convenient energy source, there is mounting concern about the damage they cause to the environment. Mining or drilling for fossil fuels damages surrounding land and habitats and increases soil erosion and flooding. Wastes pollute rivers, streams, and oceans. Spills and accidents destroy ecosystems and coastal wetlands. "The only way to avoid more fossil fuel disasters is to move aggressively away from dangerous energy sources like oil and coal,"[31] said Phillip Radford, the executive director of Greenpeace.

In addition, burning fossil fuels to produce energy releases pollutants into the atmosphere. Air pollutants such as carbon monoxide, nitrogen oxides, sulfur oxides, and hydrocarbons can harm plants, animals, and humans. These pollutants have been linked to smog, acid rain, and lung damage. One of the most significant fossil fuel emissions is carbon dioxide, a gas that traps heat in the earth's atmosphere and has been linked to global warming.

As a result, many people believe the world needs to reduce fossil fuel use and find a cleaner energy source. "We're borrowing money from China to buy oil from the Persian Gulf to burn it in ways that destroy the plants. Every bit of that's got to change. . . . The answer is to end our reliance on carbon-based fuels,"[32] said Al Gore, a former US vice president.

Low Emission Levels

Unlike fossil fuel power plants, geothermal plants do not burn fuel to generate electricity. As a result, emissions from geothermal power plants are typically very low. They release less than 1 percent of the carbon dioxide that a fossil fuel plant emits. In addition, geothermal plants emit 97 percent less acid rain–causing sulfur compounds than fossil fuel plants. Even better, geothermal binary cycle power plants have no emissions at all, because they use a closed-loop system to generate power.

Geothermal power plants do release several gases, including hydrogen sulfide and carbon dioxide, which are found naturally in geothermal reservoirs. Without geothermal development, these gases would eventually evaporate into the atmosphere, over a much slower period. To reduce emissions, many geothermal plants install scrubbers that remove the gases before they reach the atmosphere.

Lake County, California, is home to 5 geothermal power plants and has consistently ranked among the best counties nationwide for clean air since 1990. For the past 20 years, the county has complied in full with Federal Clean Air Standards and with more rigorous California standards for ozone and air pollution. In addition, Lake County was the only county in California to place in the top 10 cleanest counties for small-particulate levels, according to the American Lung Association's 2010 State of the Air Report. "It really does say a lot for the air quality we enjoy here,"[33] said Doug Gearhart, the Lake County's pollution control officer.

Greenhouse Gases

Burning fossil fuels produces carbon dioxide, a greenhouse gas that traps heat in the earth's atmosphere and has been linked to global warming. Many scientists believe that global warming poses significant threats to the environment and human health. According to the Energy Information Administration, carbon dioxide accounts for 83 percent of greenhouse gas emissions in the United States. In addition, scientists believe that burning fossil fuels over the past 150 years has resulted in more than a 25 percent increase in the amount of carbon dioxide in the atmosphere.

In contrast to fossil fuel plants, geothermal power plants emit very low levels of carbon dioxide. In addition, injecting geothermal fluids into underground reservoirs before gases release into the air reduces emissions. When Nevada's Dixie Valley geothermal power plant began re-injection, the plant's carbon dioxide emissions decreased by 39 percent.

Installation of geothermal heat pumps also reduces greenhouse gas emissions. In Langley, England, township officials installed a heat pump system at the township's water treatment plant in 2009. Before the geothermal system, the township used natural gas for heating and cooling. The new system, however, pulled thermal energy from the drinking water that was already being processed at the plant. Within a short time, township officials noticed a reduction in greenhouse gas emis-

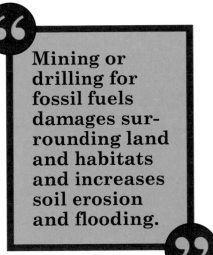

Mining or drilling for fossil fuels damages surrounding land and habitats and increases soil erosion and flooding.

sions. "We saved 31 tonnes of greenhouse gas emissions in six months,"[34] said Ryan Schmidt, the township environmental coordinator.

Waste and Fumes

Some people point out, however, that geothermal energy is not emission free or waste free. The hot water or steam of a geothermal reservoir carries naturally occurring metals, minerals, and gases to the surface. As the water or steam passes through underground rock, these potentially toxic materials can leak into the surrounding area. In addition, water used during geothermal testing can also be saturated with these minerals and chemicals. "They need to do something with the water, but it's not good to discharge it to streams or allow it to contaminate someone's drinking-water well,"[35] said Bill Mason, a groundwater hydrologist with the Oregon Department of Environmental Quality's Eugene office.

In addition, gases such as hydrogen sulfide, ammonia, methane, and carbon dioxide are released when geothermal steam reaches the surface. While these metals, minerals, and gases occur naturally, the process of obtaining geothermal energy causes them to be released much more quickly and in greater quantities than would naturally happen. At the Geysers, the steam vented at the surface contains several minerals, including hydrogen sulfide, which causes the area's rotten egg smell. Tom Tanton, general manager for renewables and hydropower at the Electric Power Research Institute, said: "Geothermal power plants tend to emit hydrogen sulfide—which is toxic at fairly low levels—and mercury. . . .

> Geothermal binary cycle power plants have no emissions at all, because they use a closed-loop system to generate power.

Whatever is not reinjected into the ground can cause local groundwater pollution. Geothermal fluids are always foul smelling—they smell like very rotten eggs due to the H_2S [hydrogen sulfide]. The fluids are highly brackish and contain high levels of heavy metals."[36]

Scrubbers can reduce emissions, cleaning the air of 99 percent of hydrogen sulfide gas. Scrubbers, however, produce a sludge waste that is high in sulfur and vanadium, a heavy metal that can be toxic in high concentrations. Additionally, geothermal systems produce sludge from dissolved solids when condensing geothermal steam. This sludge can contain silica compounds, chlorides, arsenic, mercury, nickel, and other toxic metals. To dispose of these wastes, some plants dry the sludge and ship it to hazardous waste sites. Other plants inject liquid wastes or redissolved solids into the geothermal well. Either disposal method can be costly. Alternatively, closed-loop plant designs inject steam or water back into the ground before it can be released into the air.

> " By using local wastewater to replenish steam fields, geothermal plants reduce water pollution in nearby communities. "

Geothermal systems have the ability to reduce pollution. In order to prevent a geothermal field from being depleted, geothermal systems use satellite wells around the steam well to reinject water back into the ground and recharge the geothermal resource. By using local wastewater to replenish steam fields, geothermal plants reduce water pollution in nearby communities.

Since 2003 the city of Santa Rosa, California, has partnered with the Geysers to use the city's treated wastewater to recharge the Geysers geothermal resource. Before this time, the wastewater was dumped into local tributaries. Now, miles of pipeline pump millions of gallons of wastewater each day from Santa Rosa to the Geysers. The plant distributes the wastewater around the geothermal field and recycles it into the ground, where underground rock heats it. "The wastewater injection project has been a success story. It's a great way for municipalities to deal with their wastewater, and it has given us a much steadier production profile,"[37] said Mike Rogers, geothermal manager for Calpine Corporation at the Geysers.

Land Needs of Geothermal Plants

Geothermal energy uses less land per megawatt of energy produced than other power plants. Because a geothermal plant sits directly on top of its energy source, it does not need additional land to mine coal or transport oil or gas. Advanced directional- or slant-drilling technologies allow geothermal power companies to use less land. In some fields several wells can be drilled into one location. According to the Energy Department, a geothermal power plant uses from 1 to 8 acres (0.4 to 3.2 ha) per megawatt produced. In contrast, nuclear plants use 5 to 10 acres (2 to 4 ha) per megawatt, and coal power plants use 19 acres (7.7ha).

> "
> **Drilling to shallow geothermal reserves can trigger earthquakes, but these are usually small and often go unnoticed.**
> "

In addition to requiring a smaller space, geothermal systems can share land with other operations. After the plant is built and running, surrounding fields can be used for livestock or agriculture. In Southern California's Imperial Valley, 15 geothermal power plants coexist with a rich agricultural region. In Mammoth Lakes, California, skiers and other outdoor enthusiasts may never notice the geothermal power plant that was specially designed to blend into the landscape.

Although geothermal operations require less land than other energy sources, the most promising geothermal areas are often located in scenic, wild, and protected places. "Geothermal sites often are located in protected wilderness areas that environmentalists do not want disturbed,"[38] said Robert Bradley, the president of the Institute for Energy Research in Houston. Laws protect the geothermal fields and restrict drilling in wilderness areas. Yellowstone National Park's famous geysers sit on top of a massive geothermal resource that has the potential to be the world's largest geothermal field. Yet Yellowstone's lands are free of power plants, because of regulations that put national park lands off-limits to energy development.

Risk of Earthquakes

In order to access geothermal energy in more places, developers are testing technology to drill deeper and reach hotter resources. They are also testing methods to fracture areas of hot dry rock and pump cold water underground for heating. If successful, this technology would significantly increase worldwide access to geothermal energy.

As scientists drill deeper into the ground, some people are concerned that their activity could trigger a serious earthquake. Drilling to shallow geothermal reserves can trigger earthquakes, but these are usually small and often go unnoticed. Water injected into wells to open or enlarge existing fractures can cause small tremors. "Any process that injects pressurized water at depth into rocks will cause them to fracture and possibly trigger earthquakes,"[39] said Brian Baptie, an earthquake specialist at the British Geological Society. Drilling deeper raises the risk that a more serious earthquake could occur, because seismologists know that large earthquakes originate deep within the earth.

In fact, the risk of earthquakes has put several geothermal projects on hold. In Basel, Switzerland, the government halted a geothermal project in 2006 after a series of earthquakes that scared residents and cracked buildings. In December 2009 the project was permanently shut down when a government study found that the earthquakes generated by the project were projected to cause millions of dollars of damage annually.

Residents in Northern California protested the opening of a similar geothermal project to drill deep into hot dry rock at the Geysers' geothermal field. Residents say that they have already experienced many small earthquakes because of nearby geothermal power plants, and they fear deeper wells could trigger a more serious event. "It's terrifying. What's happening to all these rocks that they're busting into a million pieces?"[40] said Susan Bartlett, an area resident. When the drilling company, AltaRock Energy, announced they were halting the project in December 2009, residents celebrated. "I'm just so relieved, because with this going on, I'm afraid one of these days it's going to knock my house off the hill,"[41] said resident Jacque Felber.

No Perfect Energy Source

Geothermal energy is one of the cleanest energy sources. Geothermal emissions contain only tiny amounts of the pollutants emitted by fossil fuels. Its footprint is small, requiring less land than most other energy sources and preserving surrounding land for complementary uses.

Despite these benefits, there are environmental drawbacks to geothermal energy. Although emissions are lower, there is still the risk that bringing underground water to the surface will contaminate local water supplies and ground with naturally occurring but potentially toxic minerals and chemicals. In addition, the process of harnessing geothermal energy has been linked to earthquakes. "It's a trade-off. You have benefits and hazards. There's no perfect technology,"[42] said Ernie Majer, a seismologist at the Lawrence Berkeley National Laboratory.

How Does Geothermal Power Affect the Environment?

> **❝** Geothermal energy helps reduce the impact of pollution on our air, water, and wildlife and its use is growing.**❞**

—Leslie Blodgett, "Celebrate Geothermal Energy as 'Mother Earth's' Energy on Earth Day 2008," Geothermal Energy Association, April 28, 2008. http://geo-energy.org.

Blodgett is the outreach director of the Geothermal Energy Association.

> **❝** We must also look better at the question of [hydrogen sulfide] pollution from geothermal production, which is a new and growing concern because of unpleasant smell and possibly for health reasons.**❞**

—Svandís Svavarsdóttir, quoted in Kate Galbraith, "Iceland's Environment Minister Discusses Geothermal Power; Aluminum Concerns," Q & A, *New York Times*, July 8, 2009. http://green.blogs.nytimes.com.

Svavarsdóttir is Iceland's minister for the environment.

Primary Source Quotes

> **"We partnered with some local communities . . . and brought in treated wastewater, which we were able to put back into the ground and regenerate steam. . . . It was a definite win-win between industry and community."**

—Larry Sessions, "Personnel Profile: Q&A with Larry Sessions," *Capitol Weekly*, November 18, 2010. www.capitolweekly.net.

Sessions is the general manager of the Geysers.

> **"I have no doubt we'll reap the rewards of our geothermal system. Our patients will be comfortable. Our system will be durable. And we will not be producing any form of pollution. Geothermal was the right way to go."**

—Jim Pernau, quoted in "New Edgerton Hospital Proves It's Cool to Be the First to Use Geothermal Heating," Edgerton Hospital press release, September 29, 2010. www.edgertonhospital.com.

Pernau is the chief executive officer of Edgerton Hospital in Edgerton, Wisconsin, which installed a geothermal heating and cooling system.

> **"The geothermal industry has a history of environmental responsibility and is committed to providing energy consumers with clean, reliable, and safe electricity."**

—Dan Jennejohn, "GEA Issue Brief: Geothermal Energy and Greenhouse Gas Emissions," Geothermal Energy Association, July 27, 2009. www.geo-energy.org.

Jennejohn is a research associate with the Geothermal Energy Association.

> **"The risk of overreaction to the risks inherent in deep geothermal projects is very real."**

—Domenico Giardini, "Geothermal Quake Risks Must Be Faced," *Nature*, December 16, 2009. www.nature.com.

Giardini is the director of the Swiss Seismological Service.

> 66 If we find the ways and means of tapping these geo-thermal energy resources and make full use of them, there would be substantially less carbon emissions, in the atmosphere of our planet. 99

—Susilo Bambang Yudhoyono, speech at the World Geothermal Congress, April 26, 2010. http://indonesia.gr.

Yudhoyono is the president of the Republic of Indonesia.

> 66 Stimulating fractures using [enhanced geothermal system] technology creates small seismic events termed 'microseismicity.' But the goal of the technology is to create events below the human detectable level. . . . Creating larger fractures that might result in seismic events large enough to be felt . . . is undesirable. 99

—Susan Petty, "Geothermal Energy—Protecting the Environment and Our Future," AltaRock Energy, June 9, 2010. http://altarockenergy.com.

Petty is the president and chief technology officer at AltaRock Energy, a geothermal energy company.

> 66 Geothermal is the cleanest of all electricity sources and is also one of the cheapest when you include all costs (capital, operating, maintenance and fuel). 99

—Ross Beaty, quoted in Ron Hera, "Oil Is Over and Geothermal Is the Way of the Future," Q & A, Business Insider, May 5, 2010. www.businessinsider.com.

Beaty is the chair and chief executive officer of the Magma Energy Corporation.

How Does Geothermal Power Affect the Environment?

- US geothermal generation annually offsets the emission of about **24 million tons** (22 million metric tons) of carbon dioxide, 200,000 tons (181,437 metric tons) of nitrogen oxides, and 110,000 tons (99,790 metric tons) of particulate matter from conventional coal-fired plants.

- A typical geothermal plant emits about **1 percent** of the sulfur dioxide, **5 percent** of the carbon dioxides, and less than **1 percent** of the nitrous oxides emitted by an equal-sized coal-fired plant.

- Combustion of bituminous coal emits about **1,980 pounds** (900kg) of carbon dioxide per megawatt-hour, whereas natural gas releases more than **660 pounds** (300kg) per megawatt-hour.

- Binary, air-cooled geothermal power plants have effectively **zero emissions**.

- Over 30 years a geothermal power facility will use **4,348 square feet** (404 sq m) of land per gigawatt-hour as compared to a coal facility, which will use 39,094 square feet (3,632 sq m) of land per gigawatt-hour.

- Scrubbers reduce hydrogen sulfide emissions from geothermal power plants by more than **99 percent**.

Carbon Dioxide Emissions by Fuel Source

Because geothermal power plants do not burn fuel, they have very low emissions. The emissions from a geothermal plant are generally natural minerals that are found underground and brought to the surface with geothermal steam or hot water. Geothermal binary cycle plants have no emissions at all, because they operate in a closed loop.

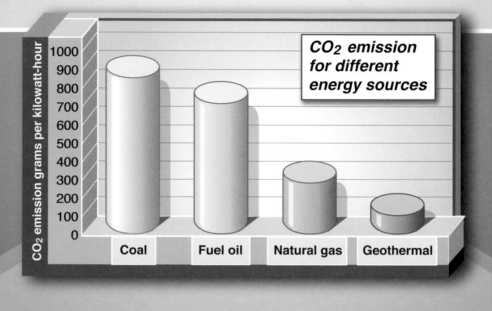

CO2 emission for different energy sources

Source: International Geothermal Association, "Geothermal: A Natural Choice," April 2010. www.geothermal-energy.org.

- Geothermal plants use **5 gallons** (19L) of fresh water per megawatt-hour, and binary air-cooled plants use no fresh water, as compared to 361 gallons (1,367L) of fresh water used per megawatt-hour by natural gas plants.

- Normal geothermal plant operation produces **less noise** than the wind rustling tree leaves.

- At the Geysers **11 million gallons** (41.6 million L) of treated wastewater from Santa Rosa are pumped daily for injection into the geothermal reservoir.

Freshwater Needs: Geothermal vs. Natural Gas Plant

Geothermal operations use surface water to cool generating equipment. In dry, arid regions, finding a suitable water source can be difficult. Even so, geothermal plants use much less surface water than comparable natural gas power plants.

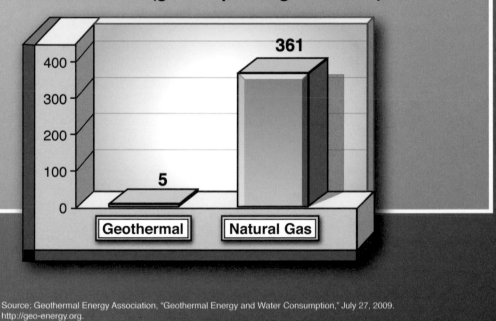

Freshwater Use (gallons per megawatt-hour)

Source: Geothermal Energy Association, "Geothermal Energy and Water Consumption," July 27, 2009. http://geo-energy.org.

- The American Lung Association estimates that power plant emissions, primarily from coal plants, result in over **30,000 yearly deaths**.

- Enhanced geothermal systems have been linked to **earthquakes** in Landau, Germany, which experienced a 2.7 magnitude quake in September 2010.

30-Year Land Use: Comparison of Alternative Energy Sources

Geothermal power uses less land than other energy sources. Because power plants are built directly on top of geothermal resources, no additional land is needed for fuel transportation or storage. In addition, once geothermal plant construction is complete, the surrounding land can be used for other activities such as agriculture, farm animal grazing, or outdoor sports.

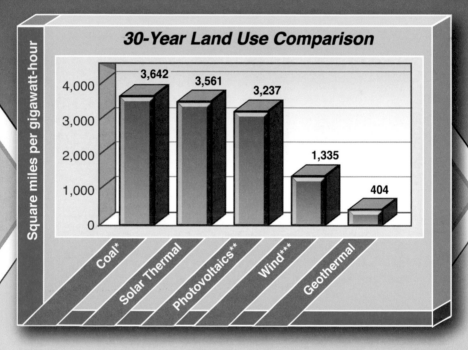

30-Year Land Use Comparison

Square miles per gigawatt-hour

Coal* 3,642
Solar Thermal 3,561
Photovoltaics** 3,237
Wind*** 1,335
Geothermal 404

*Includes mining
**Assumes central station photovoltaic project
***Land actually occupied by turbines and service roads

Source: Leslie Blodgett and Kara Slack, "Geothermal 101: Basics of Geothermal Energy Production and Use," February 15, 2009. http://geo-energy.org.

- The largest enhanced geothermal system–related seismic event, registering **3.7 magnitude**, occurred in 2003 in Cooper Basin, Australia.

- At the Geysers in California, geothermal steam contains up to **0.15 percent hydrogen sulfide** by weight before scrubbing.

What Is the Future of Geothermal Power?

66There is the potential to power millions of homes, businesses and schools from the heat of the earth. The success of geothermal power over the past 50 years gives us an incredible foundation to build a green future over the next 50 years.99

—Karl Gawell, executive director of the Geothermal Energy Association.

66If we are going to try to achieve a transformational change in this country, geothermal should be part of that recipe. It's not treated that way. It's typically forgotten.99

—Jefferson Tester, professor of sustainable energy systems at Cornell University.

Potential Future Use

Although heat from the earth is considered clean, efficient, and sustainable, widespread use of geothermal energy has been constrained by technology and expense. Despite these challenges, the future of geothermal energy is promising. Scientists are working to develop new technologies for accessing geothermal energy. In addition, experts say, the cost of geothermal energy will continue to drop as technology improves.

According to a 2010 report by ABS Energy, an independent energy market research company, the worldwide development of geothermal energy is projected to increase 78 percent between 2010 and 2015. In addition, the number of countries producing geothermal energy is expected to increase from 24 in 2009 to 36 in 2015. Some of the most promising emerging markets include Kenya, Peru, and Chile.

In the United States geothermal energy's potential is vast. According to a 2008 assessment by the US Geological Survey, the United States' geothermal potential is approximately 9,057 megawatts-electric of power potential from conventional, identified geothermal resources located in California and several other western states.

> "The number of countries producing geothermal energy is expected to increase from 24 in 2009 to 36 in 2015.

Conventional but undiscovered geothermal resources could add another 30,033 megawatts-electric of power. The biggest area of potential future use lies in unconventional, high-temperature, low-permeability areas also in western states, where new technology could potentially tap more than 517,000 megawatts-electric of power. This huge supply of energy, however, is the hardest to access and will require exploration in new areas and the development of new drilling technologies.

New Technologies to Locate Geothermal Reservoirs

One of the challenges of geothermal energy is that it can be difficult to find and expensive to develop. Most existing geothermal plants sit in areas where underground steam or hot water rises to the surface. These surface features help developers to find underground resources. Many more geothermal resources are not visible from the earth's surface and are much more difficult to find. Understandably, developers are hesitant to spend $2 million to $10 million to drill a well without knowing if they will find a useable resource. "You want to get as much extracted energy as possible for that set of wells you've drilled, to maximize the return on your investment,"[43] explained Jefferson Tester, associate director of the Cornell Centre for a Sustainable Future.

In order to tap into these blind resources, scientists have been working on a number of new technologies that allow them to locate geothermal resources more effectively. "We feel that innovative exploration approaches could help us find these blind systems that have no surface expression. We wanted to come up with a combination of technologies that would improve the hit rate,"[44] said Ed Wall, program manager of the Energy Department's geothermal technologies program.

New aerial technology uses imaging and mapping to help developers better explore geothermal fields. In this approach the company sends a plane or helicopter into the air, where geologists use instruments called magnetometers that measure differences in the earth's magnetic field to show where surface rock has been altered by hot water, but now the water sits undetected deep underground. In addition, other companies are testing high-altitude sensing technology called light detection and ranging that helps geologists look for areas where hot water deposited silica. Geologists believe that where they find these silica deposits, there is a high probability that a hot water resource still exists, buried underground.

> "New aerial technology uses imaging and mapping to help developers better explore geothermal fields."

Some new geothermal technologies are currently being used by oil and gas producers, which have years of experience drilling into the earth. Reflection seismology creates vibrations underground that measure rock density, permeability, and fracture patterns. Engineers place thousands of sensors in a series of deep holes and then create underground vibrations. The sensors record the vibrations and relay the information to instruments on the surface. Using this technology, some companies are working to create a detailed three-dimensional image of underground sites that will show the best places to drill.

Digging Deeper to Reach Geothermal Reservoirs

Another important area of research is studying ways to access geothermal resources at deeper depths. Currently, most geothermal energy comes from approximately 500 to 650 feet (150 to 200 m) below the earth's surface, where temperatures are about 43°F to 46°F (6°C to 8°C). Researchers in Norway believe it may be possible to drill deeper, to 32,800 feet (10,000m), to access hotter, more powerful geothermal resources that can reach temperatures of at least 705°F (374°C). "If we manage to produce this kind of energy it would clearly be a 'moon landing.' This is one of the few sources of energy that we really have enough of. The only thing that we need is the technology to harvest it,"[45] said Odd-Geir Lademo, a researcher at SINTEF Materials and Chemistry, a Scandinavian research organization.

> **Fracturing bedrock and injecting high-pressure water into wells causes routine minor tremors.**

To access deeper geothermal resources, drilling technology will need to improve. As drills move deeper into the earth, temperatures and pressures increase. This causes drilling equipment to weaken or melt. In fact, most electronics that control drilling equipment will only operate a short time in temperatures above 400°F (204°C).

Despite the challenges, researchers believe deep drilling holds promise. They point out that current technology allows oil companies to drill down to 16,400 feet (5,000m) in temperatures that reach 340°F (171°C). The geothermal industry may be able to adapt the knowledge and technology that the oil drilling industry already has. "Drilling technology has evolved tremendously over the past 10 years. There are test wells for oil that drill 12,000 meters into the earth. Knowledge from the oil and drilling industry may be used in the future to capture geothermal energy,"[46] said Lademo and colleague Are Lund, senior researchers at SINTEF Materials and Chemistry.

Enhanced Geothermal Systems

Scientists have also been studying ways to make geothermal energy more accessible around the world. While geothermal reservoirs are only found in certain regions, hot dry rock formations are found almost everywhere. Developing a technology to extract thermal energy from these hot rocks could expand geothermal energy access.

A technology called enhanced geothermal systems, or EGS, may help scientists create an artificial geothermal re-source in hot dry rock. Drilling thousands of feet below the earth's surface, geologists inject large amounts of high-pressure surface water into the ground and fracture the hot rock, making it more permeable so that water and steam can flow through it. As the cold water moves through the fractured rock, it heats. Then it is piped back to the surface, where it runs turbines and generates electricity. Several projects in France, Germany, Australia, and the United States are currently underway to test EGS. "The beauty of the concept is that if it works, it can work anywhere in the world,"[47] said Subir Sanyal, president of GeothermEx, a consultancy based in California.

> " Researchers hope that the IKEA project could provide a blueprint for future large-scale installations of geothermal systems in retail stores worldwide. "

EGS could significantly expand geothermal energy worldwide. According to a 2007 Massachusetts Institute of Technology report, the thermal energy about 1.9 to 6.2 miles (3 to 10 km) below the surface available in the United States is almost 140,000 times greater than the country's annual energy consumption. Even though conservative estimates project that EGS could only tap about 2 percent of that energy, the amount would still be more than needed to supply all of the United States' energy needs.

Challenges Ahead

In order for EGS technology to become widespread, it must overcome several challenges. First, the technology to drill effectively and efficiently to the required depths needs to improve. Additionally, EGS drilling costs can run more than twice the cost of conventional geothermal drilling. However, future improvements in drilling could reduce costs. "EGS technology has already been proven to work in the few areas where underground heat has been successfully extracted. And further technological improvements can be expected,"[48] said Tester, a professor of sustainable energy systems at Cornell University.

EGS has also been linked to earthquakes. Fracturing bedrock and injecting high-pressure water into wells causes routine minor tremors. Many people worry that EGS has the potential to cause a major earthquake, risking lives and property in the surrounding area. Scientists believe, however, that the risk of major earthquakes is small. "Induced seismicity has not caused the type of damaging earthquakes that people think of," explained EGS expert Allen Jelacic of the Energy Department's geothermal technologies program. "It's not going to cause your house to fall down or cause significant damage. People could learn to live with it."[49]

Coproduction with Oil and Gas

Developers may be able to access more geothermal energy through coproduction with oil and gas drilling operations. In existing oil and gas reservoirs, there is a significant amount of hot wastewater that oil and gas companies throw away. New research, however, has shown that hot wastewater can be used to generate electricity.

Using existing oil and gas wells can significantly cut the costs of geothermal production and the time needed to begin producing electricity. "The lead-time to revenue generation is about 6 months, whereas traditional geothermal can take up to five years. The wells already have known geothermal potential, and capital costs are dramatically reduced,"[50] said George Alcom Jr., a petroleum geologist and the chief executive officer of Universal GeoPower. In addition, because coproduction uses existing oil and gas facilities, there is no need to drill new wells or build transmission lines. Coproduction also benefits oil and gas drillers, because using coproduced geothermal fluids for power production extends the economic life of the well.

> A nearly limitless supply of heat and energy lies beneath the earth's surface, waiting to be tapped and developed.

A 2006 report by the National Renewable Energy Laboratory estimated that the United States has the potential to develop almost 40,000 megawatts of electricity by coproducing at oil and gas fields. According to research by Southern Methodist University, coproduction could be applied to an estimated 37,500 oil and gas wells in the Gulf Coast region alone. Several pilot programs across the country are planned, from Louisiana to California. In 2008 Ormat Technologies announced the first successful coproduction of geothermal power at an existing oil well, in a joint project with the Energy Department. "With Ormat's advancement in binary turbine technology and the increased drilling for oil and gas exploration, the U.S. is primed for additional geothermal development,"[51] said David Blackwell of Southern Methodist University.

Model for the Future—IKEA

In Centennial, Colorado, a model of geothermal's potential is scheduled to open in the fall of 2011. IKEA, the Swedish home furnishings retailer, teamed with the Energy Department's National Renewable Energy Laboratory to construct a new store that will showcase a geothermal heating system.

According to Douglass Wolfe, the store's project construction manager, the Centennial store will be the first IKEA store in the United States built with a geothermal heating and cooling system. To construct the store's heating system, developers dug 130 holes into the ground, where the temperature remains a constant 55°F (13°C) year-round. Each hole reaches 500 feet (152m) into the ground and lies beneath the store's underground parking garage.

Scientists from the National Renewable Energy Laboratory have been collecting data during construction and will continue monitoring the store's system once it opens. "We're trying to collect data on how it actually performs. . . . By collecting actual live data on the performance of systems, you have better insight on what needs to be improved. We'll be able to say with confidence, 'if you do it this way, it will work this well,'"[52] said Erin Anderson, a senior geothermal analyst. Researchers hope that the IKEA project could provide a blueprint for future large-scale installations of geothermal systems in retail stores worldwide.

Although the geothermal system is expensive, developers expect to recoup that investment through lower energy bills. "The cost of digging 130 holes 500 feet deep and filling them with loops of pipe will be paid back in energy savings within a reasonable time. And we expect the system to last as long as the building lasts,"[53] said Wolfe. Because geothermal systems use 25 to 50 percent less electricity than conventional heating or cooling systems, widespread use of these systems could lead to a potential savings of several billion dollars each year in energy costs.

Advances Lead to New Possibilities

Today the world faces many uncertainties about future energy supplies. As more countries become industrialized, the demand for energy will continue to increase. At the same time, the supply of fossil fuels is finite. As the supply of these fuels dwindles and they become more expensive, alternative energy sources such as geothermal energy are likely to become more desirable.

A nearly limitless supply of heat and energy lies beneath the earth's surface, waiting to be tapped and developed. To date, geothermal energy development has been limited by technology and expense. However, advances in exploration, drilling technology, and techniques may soon allow the world to access geothermal energy more efficiently, effectively, and affordably. "This current geothermal renaissance is leveraging all these new technologies to go places we haven't been before,"[54] said Brigette Martini, a senior staff geologist at Ormat Technologies. As more people turn to geothermal for energy, it has the potential to play a significant role in a cleaner, more sustainable energy future.

Primary Source Quotes*

What Is the Future of Geothermal Power?

66 In the past 50 years, geothermal energy has blazed the trail for renewable power in this nation. But this industry has not peaked; the growth potential for geothermal energy is incredible. 99

—Karl Gawell, "Geothermal's Golden Year: After 50 Years Geothermal Energy Still Growing," Geothermal Energy Association, August 23, 2010. http://geo-energy.org.

Gawell is the executive director of the Geothermal Energy Association.

....

66 Advancing geothermal technology and establishing energy policy that helps promote development of geothermal and other renewable power projects is important as the United States looks for additional green sources of energy. 99

—Peter Christman, "Geothermal Energy: International Market Update 2010 to Be Unveiled at Geothermal Energy Association Global Geothermal Showcase and Forum in Washington, DC," Geothermal Energy Association, May 10, 2010. http://geoenergy.org.

Christman is the president of Pratt & Whitney Power Systems.

....

* Editor's Note: While the definition of a primary source can be narrowly or broadly defined, for the purposes of Compact Research, a primary source consists of: 1) results of original research presented by an organization or researcher; 2) eyewitness accounts of events, personal experience, or work experience; 3) first-person editorials offering pundits' opinions; 4) government officials presenting political plans and/or policies; 5) representatives of organizations presenting testimony or policy.

Primary Source Quotes

"While exploration and drilling technologies and practices are somewhat effective, opportunities to significantly improve upon conventional technology as well as develop more advanced 'revolutionary' exploration and drilling tools exist."

—Dan Jennejohn, *Research and Development in Geothermal Exploration and Drilling*, Geothermal Energy Association, December 2009. http://geo-energy.org.

Jennejohn is a research associate with the Geothermal Energy Association.

"The United States is blessed with vast geothermal energy resources, which hold enormous potential to heat our homes and power our economy. These investments in America's technological innovation will allow us to capture more of this clean, carbon free energy at a lower cost than ever before."

—Steven Chu, "Department of Energy Awards $338 Million to Accelerate Domestic Geothermal Energy," US Department of Energy, October 29, 2009. www.energy.gov.

Chu is secretary of the US Department of Energy.

"Oil and gas have found that a lot of their waste products have a significant amount of hot water. There are geothermal processes like binary or flash that allow us to take their wastewater and generate power from that."

—Larry Sessions, "Personnel Profile: Q&A with Larry Sessions," *Capitol Weekly*, November 18, 2010. www.capitolweekly.net.

Sessions is the general manager of the Geysers.

❝Drilling methods are improving, the technology for transferring energy from deep underground is advancing and economic conditions may soon make it more cost-competitive.❞

—Steve Mabee, "UMass Amherst and Connecticut Geologists Position New England for Success in the Geothermal Power Era," University of Massachusetts–Amherst, November 8, 2010. www.umass.edu.

Mabee is the state geologist of Massachusetts.

❝Commercialization of the EGS [enhanced geothermal systems] technology is going to be a long-term proposition—10 years and possibly longer—that requires sustained investment.❞

—Frank Monastero, "Q&A with GRC President Frank Monastero," *GRC Bulletin*, September/October 2009. www.geothermal.org.

Monastero is the president of the Geothermal Resources Council.

❝We believe EGS [enhanced geothermal system] is a clean, renewable energy alternative capable of reducing our country's dependence on fossil fuels and creating a more sustainable future.❞

—Donald O'Shei, "AltaRock Energy to Suspend Demonstration Project Drilling While Continuing EGS Development," AltaRock Energy, September 2, 2009. http://altarockenergy.com.

O'Shei is the chief executive officer of AltaRock Energy.

What Is the Future of Geothermal Power?

- According to the National Renewable Energy Laboratory, the processed water from oil and gas wells has the potential to generate almost **44,000 megawatts** of geothermal power.

- Geothermal companies hope coproduction technology will allow them to produce electricity from the **25 billion barrels** of geothermally heated wastewater from oil wells in the United States.

- According to the US Geological Survey, there are approximately **30,000 megawatts** of untapped conventional geothermal power in the western United States, enough to power California.

- The worldwide development of geothermal energy is projected to increase **78 percent** between 2010 and 2015.

- Enhanced geothermal systems can produce **thermal energy** by circulating water through stimulated regions of rock at depths of 1.8 to 3 miles (3 to 5 km).

- Current geothermal systems rarely require drilling beyond **1.8 miles** (3 km), but deep drilling research is attempting to reach depths of up to **6.2 miles** (10 km).

How Enhanced Geothermal Systems Work

Scientists are excited about the future possibilities of enhanced geothermal systems (EGS). Using EGS technology, scientists drill into hot, dry rock and insert water to create an artificial geothermal reservoir. Because hot, dry rock is found in most places, scientists believe that EGS may make geothermal power more accessible around the world.

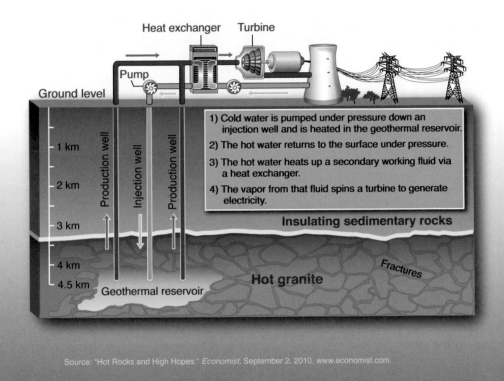

Source: "Hot Rocks and High Hopes," *Economist*, September 2, 2010. www.economist.com.

- A panel of geothermal experts convened by the Massachusetts Institute of Technology estimated that enhanced geothermal systems could provide **10 percent** of US electricity by 2050.

- The exploratory phases of a geothermal project have a **75 percent** chance of failure rate, due to the high temperatures found in geothermal reservoirs and uncertainties regarding reservoir geology.

US Geothermal Potential by Category

According to a US Geological Survey, there is enormous potential to capture geothermal energy in hot, dry rock. Scientists estimate that enhanced geothermal systems (EGS) technology could generate 93 percent of the US geothermal potential by creating artificial geothermal reservoirs in areas with hot, dry rock formations.

Future Geothermal Potential by Resource Category

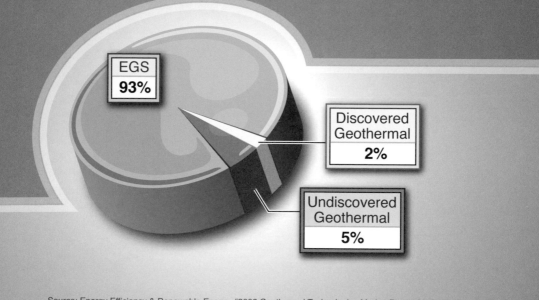

Source: Energy Efficiency & Renewable Energy, "2008 Geothermal Technologies Market Report," July 2009. www.eere.energy.gov.

- The International Geothermal Association says that there could be around **80,000 megawatts** of enhanced geothermal systems projects completed globally in the next 40 years.

- The cost of enhanced geothermal systems may impact future use. Drilling an enhanced geothermal system initial borehole is projected to cost from **$12 million to $20 million**, approximately 2 to 5 times greater than the cost to drill an oil or gas well of comparable depth.

Coproduction Potential

Most geothermal power resources in the United States are found in the western states. Geothermal coproduction with oil and gas wells offers the possibility of expanding geothermal access to regions in the central, southeastern, and northeastern states. In existing oil and gas reservoirs, a significant amount of hot wastewater is created and thrown away. New research has shown that hot wastewater can be used to generate electricity.

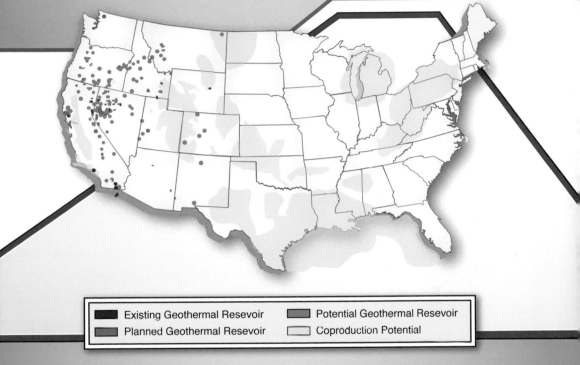

Existing Geothermal Resevoir   Potential Geothermal Resevoir
Planned Geothermal Resevoir   Coproduction Potential

Note: A geothermal reservoir is underground and contains water or steam.

Source: Energy Efficiency & Renewable Energy, "2008 Geothermal Technologies Market Report," July 2009. www.eere.energy.gov.

- According to the Western Governors Association, development of near-term geothermal potential of 5,600 MW of geothermal energy would result in the creation of almost **100,000 jobs**.

Key People and Advocacy Groups

AltaRock Energy: AltaRock Energy is a renewable energy development company focused on the research and development of enhanced geothermal systems.

Calpine Corporation: Calpine is the largest geothermal power producer in the United States. The company owns and operates 15 power plants at the Geysers with a net generating capacity of about 725 megawatts of electricity—enough to power 725,000 homes, or a city the size of San Francisco.

Energy Efficiency and Renewable Energy: Part of the US Department of Energy, the Energy Efficiency and Renewable Energy's Geothermal Technologies Program works with industry to establish geothermal energy as an economically competitive contributor to the US energy supply.

Karl Gawell: Gawell has been the executive director of the Geothermal Energy Association since 1997 and is a well-known figure in the geothermal industry.

Geothermal Education Office: Funded by the US Department of Energy and the geothermal industry, the education office promotes public understanding about geothermal resources and distributes educational materials.

Geothermal Energy Association: The Geothermal Energy Association is a US trade group that encourages research to improve geothermal technologies, develops geothermal resources worldwide for electrical power generation and direct-heat uses, and provides a forum for discussion of geothermal issues.

Geothermal Resources Council: The Geothermal Resources Council is an association of geothermal professionals and companies from around the world that promotes development, outreach, information, and technology transfer among its members.

The Geysers: The Geysers, comprising 45 square miles (117 sq km) along the Sonoma and Lake County border in Northern California, is the largest complex of geothermal power plants in the world.

International Geothermal Association: The International Geothermal Association is an organization that encourages research, development, and utilization of geothermal resources worldwide.

National Renewable Energy Lab: The National Renewable Energy Laboratory is the nation's primary laboratory for renewable energy and energy efficiency research and development. The laboratory's researchers estimated the US geothermal resource in a 2006 report.

Jefferson Tester: Tester is an expert on geothermal energy and is currently a professor of sustainable energy systems at Cornell University. He is a member of several energy advisory boards, including the National Renewable Energy Laboratory.

US Energy Association: The US Energy Association is the US member committee of the World Energy Council. The association represents the broad interests of the US energy sector by increasing the understanding of energy issues, both domestically and internationally.

Chronology

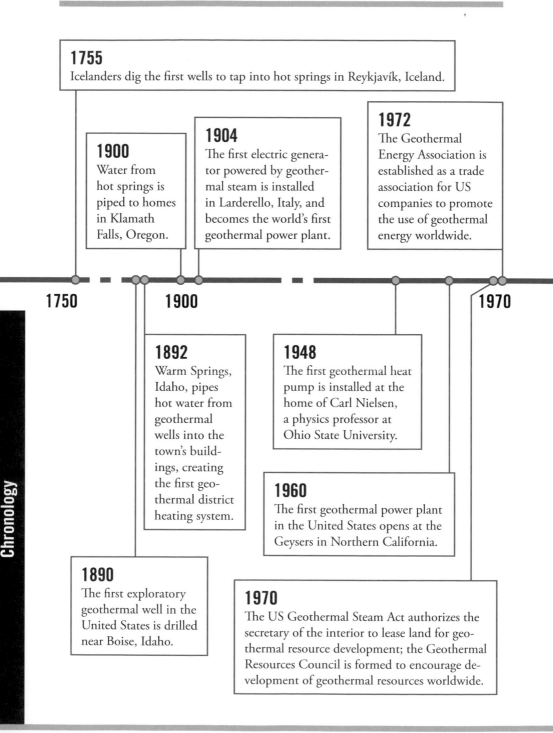

1755
Icelanders dig the first wells to tap into hot springs in Reykjavík, Iceland.

1900
Water from hot springs is piped to homes in Klamath Falls, Oregon.

1904
The first electric generator powered by geothermal steam is installed in Larderello, Italy, and becomes the world's first geothermal power plant.

1972
The Geothermal Energy Association is established as a trade association for US companies to promote the use of geothermal energy worldwide.

1750

1900

1970

1892
Warm Springs, Idaho, pipes hot water from geothermal wells into the town's buildings, creating the first geothermal district heating system.

1948
The first geothermal heat pump is installed at the home of Carl Nielsen, a physics professor at Ohio State University.

1960
The first geothermal power plant in the United States opens at the Geysers in Northern California.

1890
The first exploratory geothermal well in the United States is drilled near Boise, Idaho.

1970
The US Geothermal Steam Act authorizes the secretary of the interior to lease land for geothermal resource development; the Geothermal Resources Council is formed to encourage development of geothermal resources worldwide.

Chronology

1979

The first electrical development of a water-dominated geothermal resource occurs at the East Mesa field in the Imperial Valley in California.

2010

Researchers uncover the largest geothermal hot spot in the eastern United States, finding that West Virginia sits atop several hot patches of earth, some as warm as 392°F (200°C) and as shallow as 3 miles (5km).

1981

Ormat Technologies successfully demonstrates binary technology in the Imperial Valley in California and establishes the technical feasibility of larger-scale commercial binary power plants.

2009

An enhanced geothermal systems drilling site is permanently shut down in Basel, Switzerland, after drilling activity triggers earthquakes. A similar project is also shut down in Northern California.

1980 **1995** **2010**

1974

The first borehole to test the feasibility of enhanced geothermal systems is drilled at the Fenton Hill test site in New Mexico.

1988

After several years of expansion, production at the Geysers peaks at about 2,000 megawatts. Production declines as steam pressure decreases.

2000

The US Department of Energy initiates its GeoPowering the West program to encourage development of geothermal resources in the western United States.

2006

Well drilling for an enhanced geothermal system project in Basel, Switzerland, triggers an earthquake that registers 3.4 on the Richter scale.

1973

The price of oil sharply increases as some members of the Organization of the Petroleum Exporting Countries place an embargo on oil shipments to the United States.

2008

Scientists with the US Geological Survey complete an assessment of the United States' geothermal resources, noting that the electric power generation for identified systems is 9,057 megawatts-electric, 30,033 megawatts-electric for undiscovered resources, and another estimated 517,800 megawatts-electric from enhanced geothermal systems.

Related Organizations

Energy Efficiency and Renewable Energy

1000 Independence Ave. SW
Washington, DC 20585
phone: (877) 337-3463
website: www.eere.energy.gov

Part of the US Department of Energy, the Energy Efficiency and Renewable Energy's Geothermal Technologies Program works with industry to establish geothermal energy as an economically competitive contributor to the US energy supply.

Geothermal Education Office

664 Hilary Dr.
Tiburon, CA 94920
fax: (415) 435-7737
e-mail: geo@marin.org • website: http://geothermal.marin.org

Funded by the US Department of Energy and the geothermal industry, the education office promotes public understanding about geothermal resources and distributes educational materials.

Geothermal Energy Association

209 Pennsylvania Ave. SE
Washington, DC 20003
phone: (202) 454-5261 • fax: (202) 454-5265
website: www.geo-energy.org

The Geothermal Energy Association is a US trade group that encourages research to improve geothermal technologies, develops geothermal resources worldwide for electrical power generation and direct-heat uses, and provides a forum for discussion of geothermal issues.

Geothermal Program

1516 Ninth St., MS-43
Sacramento, CA 95814
phone: (916) 653-7551
website: www.energy.ca.gov

The California Energy Commission is the state's primary energy policy and planning agency. The commission's Geothermal Program promotes the research, development, demonstration, and commercialization of California's enormous geothermal energy sources.

Geothermal Resources Council

PO Box 1350
2001 Second St., Suite 5
Davis, CA 95617
phone: (530) 758-2360 • fax: (530) 758-2839
e-mail: crobinson@geothermal.org • website: www.geothermal.org

The Geothermal Resources Council is an association of geothermal professionals and companies from around the world that promotes development, outreach, information, and technology transfer among its members.

International Geothermal Association (IGA)

Sundurlandsbraut 48
108 Reykjavík, Iceland
phone: +354 588 4437 • fax: +354 588 4431
email: iga@samorka.is • website: http://iga.igg.cnr.it

The IGA is an organization that encourages research, development, and utilization of geothermal resources worldwide through the compilation, publication, and dissemination of scientific and technical data and information, both within the community of geothermal specialists and between geothermal specialists and the public.

International Ground Source Heat Pump Association (IGSHPA)

374 Cordell S.
Stillwater, OK 74078
phone: (405) 744-5175 • fax: (405) 744-5283
e-mail: igshpa@okstate.edu • website: www.igshpa.okstate.edu

The IGSHPA is a nonprofit, member-driven organization established in 1987 to advance geothermal heat pump technology on local, state, national, and international levels.

National Geothermal Collaborative (NGC)

1580 Lincoln St., Suite 1080
Denver, CO 80203
phone: (303) 861-1500
e-mail: info@geocollaborative.org • website: www.geocollaborative.org

The NGC's broad objectives are to provide a forum for identifying and discussing issues that affect the use of geothermal power, to catalyze problem-solving actions to address key issues, and to build consensus among varied stakeholder groups.

National Renewable Energy Lab (NREL)

1617 Cole Blvd.
Golden, CO 80401
phone: (303) 275-3000
website: www.nrel.gov

The NREL is the nation's primary laboratory for renewable energy and energy efficiency research and development.

World Energy Council (WEC)

5th Floor, Regency House
1-4 Warwick St.
London W1B 5LT
United Kingdom
phone: +44 20 7734 5996
fax: +44 20 7734 5926
e-mail: info@worldenergy.org • website: www.worldenergy.org

The WEC has almost 100 member countries and covers all types of energy, including coal, oil, natural gas, nuclear, hydro, and renewables. The WEC collects data and promotes energy research, holds workshops and seminars, and collaborates with other energy organizations.

For Further Research

Books

Leslie Blodgett and Kara Slack, *Geothermal 101: Basics of Geothermal Energy Production and Use*. Washington, DC: Geothermal Energy Association, 2009.

Bruce D. Green and R. Gerald Nix, *Geothermal—the Energy Under Our Feet*. Golden, CO: National Renewable Energy Laboratory, 2006.

Ernst Huenges and Patrick Ledru, *Geothermal Energy Systems: Exploration, Development, and Utilization*. Hoboken, NJ: Wiley, 2010.

Neil Morris, *Geothermal Power*. Mankato, MN: Smart Apple Media, 2009.

Dana Meachen Rau, *Alternative Energy: Beyond Fossil Fuels*. Mankato, MN: Compass Point, 2010.

Fiona Reynoldson, *Understanding Geothermal Energy and Bioenergy*. New York: Gareth Stephens, 2010.

Lorraine Savage, *Geothermal Power*. Farmington Hills, MI: Greenhaven, 2007.

John Tabak, *Solar and Geothermal Energy*. New York: Facts On File, 2009.

Alan Watchel, *Geothermal Energy*. New York: Chelsea House, 2010.

Periodicals

David Biello, "Deep Geothermal: The Untapped Energy Source," *Yale Environment 360*, October 23, 2008.

———, "Hot Rocks: Tapping an Underutilized Renewable Resource," *Scientific American*, January 23, 2007.

Leslie Blodgett, "Oil and Gas Coproduction Expands Geothermal Power Possibilities," *Renewable Energy World*, July 9, 2010.

James Glanz, "Deep in Bedrock, Clean Energy, and Quake Fears," *New York Times*, June 23, 2009.

"Hot Rocks and High Hopes," *Economist*, September 2, 2010.

Christopher Mims, "One Hot Island: Iceland's Renewable Geothermal Power," *Scientific American*, October 20, 2008.

Denise Rockenstein, "Geysers Celebrate 50 Years," *Lake County Record-Bee* (CA), September 13, 2010.

Cassandra Sweet, "Not Just a Lot of Hot Air," *Wall Street Journal*, September 13, 2010.

Internet Sources

Geothermal Education Office, "Geothermal Energy Facts," www.geo thermal.marin.org/pwrheat.html.

Geothermal Energy Association, "Green Jobs Through Geothermal Energy," October 2010. http://geo-energy.org/pdf/reports/GreenJobs_Through_Geothermal_Energy_Final_Oct2010.pdf.

———, "Research and Development in Geothermal Exploration and Drilling," December 2009. http://geo-energy.org/pdf/reports/Explo rationandDrillingR&D.pdf.

———, "US Geothermal Power Production and Development Update," April 2010. http://geo-energy.org/pdf/reports/April_2010_US_Geo thermal_Industry_Update_Final.pdf.

"IKEA Geothermal System Could Inform Others," National Renewable Energy Laboratory, August 19, 2010. www.nrel.gov/features/20100819_geothermal.html.

Massachusetts Institute of Technology, *The Future of Geothermal Energy*. Idaho Falls: Idaho National Laboratory, 2006. http://geothermal.inel.gov/publications/future_of_geothermal_energy.pdf.

Source Notes

Overview

1. Quoted in Christopher Mims, "One Hot Island: Iceland's Renewable Geothermal Power," *Scientific American*, October 20, 2008. www.scientificamerican.com.
2. Wendell A. Duffield and John H. Sass, "Geothermal Energy—Clean Power from the Earth's Heat," *US Geological Survey Circular 1249*, 2003. http://pubs.usgs.gov.
3. Quoted in "Claims Geothermal Energy Is Solution for Future," *Eurasia Review*, September 7, 2010. www.eurasiareview.com.
4. Quoted in "Chemical Engineer: Geothermal Is Undervalued US Energy Source," Massachusetts Institute of Technology, February 13, 2007. http://web.mit.edu.
5. Quoted in David Biello, "Hot Rocks: Tapping an Underutilized Renewable Resource," *Scientific American*, January 23, 2007. www.scientificamerican.com.
6. Quoted in Biello, "Hot Rocks."
7. Quoted in Biello, "Hot Rocks."
8. Quoted in Biello, "Hot Rocks."
9. Quoted in David Biello, "Drilling for Hot Rocks: Google Sinks Cash into Advanced Geothermal Technology," *Scientific American*, August 20, 2008. www.scientificamerican.com.

Can Geothermal Power Supply the World's Energy Needs?

10. Quoted in Julie Taylor, "Making Power in NZ from the Ground Up," *Daily Post* (Rotorura, New Zealand), December 3, 2010. www.rotoruadailypost.co.nz.
11. Leslie Blodgett, "The Recognition of Geothermal Energy's Benefit Expanding in 2010 and Beyond," *Alternative Power Construction*, October 5, 2010. http://altpowerconstruction.com.
12. Quoted in Denise Rockenstein, "Geysers Celebrate 50 Years," *Lake County Record-Bee* (CA), September 13, 2010. www.record-bee.com.
13. Quoted in Steve Hart, "The Geysers at 50," *Press Democrat*, October 3, 2010. www.pressdemocrat.com.
14. Quoted in Andy Harvey, "Geothermal Water Keeps School Warm," Keloland.com, March 3, 2006. http://keloland.com.
15. Quoted in Charles Q. Choi, "The Energy Debates: Geothermal Energy," LiveScience, December 10, 2008. www.livescience.com.
16. Quoted in "Making Geothermal More Productive," University of Utah, September 8, 2009. www.unews.utah.edu.
17. Quoted in "Beyond Fossil Fuels: Lucien Bronicki on Geothermal Energy," *Scientific American*, April 30, 2009. www.scientificamerican.com.
18. Quoted in Hillary Brenhouse, "Indonesia Seeks to Tap Its Huge Geothermal Reserves," *New York Times*, July 26, 2010. www.nytimes.com.
19. Quoted in Christopher Mims, "Can Geothermal Power Compete with Coal on Price?," *Scientific American*, March 2, 2009. www.scientificamerican.com.
20. Massachusetts Institute of Technology, *The Future of Geothermal Energy*. Idaho Falls: Idaho National Laboratory, 2006. http://geothermal.inel.gov.

Can Geothermal Power Reduce Dependence on Fossil Fuels?

21. Quoted in Onell R. Soto, "Just Color San Diego Algae Green in Research," *San Diego Union-Tribune*, November 12, 2010. www.signonsandiego.com.
22. Quoted in "Obama: End Dependence on Fossil Fuels," MSNBC.com, June 2, 2010. www.msnbc.msn.com.
23. Quoted in "Geothermal Mapping Project Reveals Large, Green Energy Source in Coal Country," Physorg.com, October 5, 2010. www.physorg.com.
24. Quoted in Laura Lundquist, "Bill Would Create Tax Break for Geothermal Projects in US," *Twin Falls (ID) Times-News Magic Valley*, September 17, 2010. www.magicvalley.com.
25. Quoted in Brenhouse, "Indonesia Seeks to Tap Its Huge Geothermal Reserves."
26. "The Hidden Costs of Fossil Fuels," Union of Concerned Scientists, October 29, 2002. www.ucsusa.org.
27. Quoted in Mitchell Andersen, "Why Is Canada Freezing Out Geothermal Power?," *Tyee* (BC), November 19, 2010. http://thetyee.ca.
28. Quoted in Jennifer Keefe, "Geothermal Energy Use Gaining Steam in the US" Fosters.com, October 17, 2010. www.fosters.com.
29. Quoted in Rebecca Jacobs, "Hawaii Takes a Close Look at Geothermal Energy," *Indian Country Today*, December 7, 2010. www.indiancountrytoday.com.
30. Quoted in "Hot Rocks and High Hopes," *Economist*, September 2, 2010. www.economist.com.

How Does Geothermal Power Affect the Environment?

31. Phillip Radford, "President Obama: Give Us Our Future Back," Green-peace, June 15, 2010. www.greenpeace.org.
32. Al Gore, "Al Gore's Speech at Constitution Hall in Washington," National Public Radio, July 17, 2008. www.npr.org.
33. Quoted in Elizabeth Larson, "'State of the Air Report' Gives Lake County High Grades for Another Year," *Lake County News* (CA), April 30, 2010. http://lakeconews.com.
34. Quoted in "Plant's New System Saves Gas and Cash," *Aldergrove (BC) Star*, November 4, 2010. www.bclocalnews.com.
35. Quoted in "Oregon Streamlining Permit for Geothermal Drilling," *Salem (OR) Statesman Journal*, December 7, 2010. www.statesmanjournal.com.
36. Quoted in James M. Taylor, "Geothermal Power Would Harm California, Claims Lawsuit," *Environment & Climate News*, July 1, 2004. www.heartland.org.
37. Quoted in Hart, "The Geysers at 50."
38. Quoted in Taylor, "Geothermal Power Would Harm California, Claims Lawsuit."
39. Quoted in Shanta Barley, "Geothermal Power Quakes Find Defenders," *New Scientist*, September 16, 2009. www.newscientist.com.
40. Quoted in James Glanz, "Deep in Bedrock, Clean Energy and Quake Fears," *New York Times*, June 23, 2009. www.nytimes.com.
41. Quoted in James Glanz, "Geothermal Project in California Is Shut Down," *New York Times*, December 11, 2009. www.nytimes.com.
42. Quoted in "Hot Rocks and High Hopes."

What Is the Future of Geothermal Power?

43. Quoted in "Hot Rocks and High

Hopes."

44. Quoted in Cassandra Sweet, "Not Just a Lot of Hot Air," *Wall Street Journal*, September 13, 2010. http://online.wsj.com.

45. Quoted in "Drilling 10,000 M Deep Geothermal Wells," *Renewable Energy Focus*, September 15, 2010. www.renewableenergyfocus.com.

46. Quoted in "Drilling 10,000 m Deep Geothermal Wells."

47. Quoted in "Hot Rocks and High Hopes."

48. Quoted in "MIT-Led Panel Backs 'Heat Mining' as Key US Energy Source," *MIT News*, January 22, 2007. http://web.mit.edu.

49. Quoted in David Biello, "Deep Geothermal: The Untapped Energy Source," *Yale Environment 360*, October 23, 2008. http://e360.yale.edu.

50. Quoted in Leslie Blodgett, "Oil and Gas Coproduction Expands Geothermal Power Possibilities," *Renewable Energy World*, July 9, 2010. www.renewableenergyworld.com.

51. Quoted in "Geothermal Energy Improves US Oil Recovery," Ormat Technologies, October 19, 2008. http://investor.ormat.com.

52. Quoted in "IKEA Geothermal System Could Inform Others," National Renewable Energy Laboratory, August 19, 2010. www.nrel.gov.

53. Quoted in "IKEA Geothermal System Could Inform Others."

54. Quoted in Sweet, "Not Just a Lot of Hot Air."

List of Illustrations

List of Illustrations

Index

Note: Boldface page numbers indicate illustrations.

Index

About the Author

Carla Mooney is the author of many books for young adults and children. She lives in Pittsburgh, Pennsylvania, with her husband and three children.